AND PROTECTION PAPER

119

Quarantine for seed

Edited by
S.B. Mathur
Danish Government Institute of Seed Pathology for Developing Countries
Hellerup, Copenhagen, Denmark
and
H.K. Manandhar
Central Division of Plant Pathology
Nepal Agricultural Research Council
Khumaltar, Latipur, Nepal

Proceedings of the Workshop on Quarantine for Seed in the Near East, held at the International Center for Agricultural Research in the Dry Areas (ICARDA), Aleppo, Syrian Arab Republic, 2 to 9 November 1991

Food and Agriculture Organization of the United Nations

Rome, 1993

M-13
ISBN 92-5-103324-2

Participants and Instructors of the Workshop on Quarantine for Seed in the Near East

Cover-photo: Distribution map of plant diseases, *Ascochyta fabae* in *Vicia faba* (top left); wheat seed infested with nematode galls, *Anguina tritici* (top right); seed health testing of sorghum by blotter method (bottom left); and seed health testing of chickpea on modified Czapek Dox agar (bottom right) (Photo: M. Ragab)

Preface

Due to their nature, seeds are easy to move over unlimited distances. They are also easy to store in viable conditions for short or lengthy periods of time. They can harbour organisms of quarantine significance such as fungi, bacteria, viruses, nematodes and insects, and can be contaminated by unwanted weed seeds. Well documented cases tell the story of the spread of devastating pests from one area, country or continent to another: Bunt, flag and loose smut of wheat, bacterial canker of tomato, bacterial blights of bean and soybean, and barley stripe mosaic are just a few examples from an ever-increasing list of seed-borne pests which have spread and have been intercepted over wide distances.

Introductions of high-yielding cultivars, pest-resistant varieties and germplasm, mostly as seeds, are used as one of the major inputs to ameliorate productivity and quality of agricultural produce required to bridge the increasing gab between food production and consumption in the Near East Region.

Such introductions are made on a small scale by farmers and research institutions, or in bulk consignments imported by the commercial/development sector. However, in most cases, this situation leads to the introduction of seed-borne pests, thereby contributing to crop losses.

The Workshop on "Quarantine for Seed in the Near East" held from 2 to 9 November 1991 in Aleppo, Syria, focused on this issue and furnished ways and means gained from national, regional and international experiences to contain/minimize the spread of quarantine pests through seed introductions.

It is hoped that this publication will have an impact on quarantine measures related to seed introduction and concequently on the promotion of crop health and production.

Atif Y. Bukhari
Assistant Director-General
and Regional Representative
for the Near East

Contents

Acknowledgement

The editors wish to express their gratitude to Ms Lizzi Courtney, Secretary of the Danish Government Institute of Seed Pathology for Developing Countries, for the excellent secretarial help in correcting and typing of manuscripts and in finalizing the layout of these ICARDA proceedings.

Introduction
Opening Session
and
Key Note Lecture

Introduction

The workshop was held during the period, 2 to 9 November 1991, in Aleppo, Syria. It was organized jointly by the Near East Regional Office of the Food and Agriculture Organization of the United Nations (FAO), the International Center for Agricultural Research in the Dry Areas (ICARDA) and the Danish Government Institute of Seed Pathology for Developing Countries (DGISP). The workshop was attended by participants from Algeria, Cyprus, Egypt, Jordan, Libya, Morocco, Oman, Pakistan, Sudan, Syria, Turkey and Yemen.

Opening Session

In the opening session, Dr. Marlene Diekmann welcomed the workshop participants on behalf of the Organizing Committee. She expressed thanks to the organizations supporting the workshop and the hope that some of the problems that still exist in quarantine could be solved by cooperation among the countries of the West Asian and North African regions and with the other institutions represented at the workshop.

Dr. Mahmoud Taher, on behalf of FAO, welcomed the participants and expressed his thanks to ICARDA for hosting the workshop; to the governments of Algeria, Cyprus, Egypt, Jordan, Libya, Morocco, Oman, Pakistan, Sudan, Syria, Tunisia and Yemen for their interest in the workshop; and to CTA, EPPO, GTZ, ICRISAT and INRA for supporting and participating in the workshop. Dr. Taher emphasized the importance of quarantine for seed in minimizing the introduction and spread of seed-borne pests from one area/region/continent to another and stressed the need for regional and international cooperation and coordination to ensure safe movement of seed. He also brought to the attention of the workshop, the current efforts of FAO, international research centres, regional plant protection organizations and research institutions in ensuring safe movement of seed/germplasm, which have led to the enactment of regulations on quarantine for seed; the establishment of quarantine facilities at international research centres; and the development of reliable detection techniques and production of guidelines on the safe movement of germplasm. Dr. Taher pointed out that the Near East countries are trying to bridge the present gap between food production and nutritional requirements for their populations by the introduction of high-yielding cultivars, pest-resistant varieties and/or germplasm to improve the quality and quantity of food production. As the level of quarantine for seed in most of these countries seems to be unsatisfactory, these introductions are often coupled with the occurrence of seed-borne pests which lead to crop losses, under-production and the implementation of expensive plant protection measures. Dr. Taher concluded that the objectives of the workshop were to assess the health status of seed in the Near East as well as the available capabilities on quarantine for seed; to provide up-to-date knowledge on seed-borne pests and their detection; to list essential facilities required for seed health testing for quarantine; to prepare a tentative list of seed-borne pests of quarantine significance; and to provide demonstrations of techniques used in detection of pests in seed.

Dr. Mohwar Saxena conveyed the good wishes of Dr. N.R. Fadda, Director General of ICARDA, and explained ICARDA's mandate to improve food production in the countries of West Asia and North Africa. He added that the exchange of germplasm and breeding material is an important factor in crop improvement, and measures should be taken to minimize the risk of spreading pests with this material. Finally, he expressed the hope that the workshop would prove fruitful.

International Spread and Interception of Seed-borne Pests

S.B. MATHUR* and MARLENE DIEKMANN**

*Danish Government Institute of Seed Pathology for Developing Countries, Ryvangs Allé 78, DK-2900 Hellerup, Denmark; **International Center for Agricultural Research in the Dry Areas, ICARDA, P.O. Box 5466, Aleppo, Syria

Introduction

Plant pests have been spreading to new areas ever since plant propagated material started moving from one location to another. In his book on seed pathology Neergaard (1979) gave a rather comprehensive account of spread of diseases from one country to another, including interceptions. He cited 18 examples of fungi, 5 of bacteria and 2 of viruses for introduction of diseases to new areas and 23 fungi, 3 bacteria, 2 viruses and 2 nematodes for interceptions. The lists compiled by Neergaard can be updated by adding reports on spread and interceptions published in literature, especially the Review of Plant Pathology and FAO Plant Protection Bulletin.

The published literature contains names of diseases which are known to cause heavy losses and which have a high rate of multiplication and spread. Plant protection scientists know how difficult and expensive it is to circumvent and eradicate an exotic disease once it has been introduced. Problems connected with introduction of new genetic material include aspects such as introduction of new dangerous races and possibilities of exposure of incoming material to severe attacks by local races of pests. These aspects have been the topics of discussions at national and international levels.

Pests are also disseminated in other ways. Live Colorado beetles, for example, were found in imported car spare parts (Bartlett, 1990). Long-distance spread of spores by wind occurred when sugarcane rust spread from Cameroon to the Dominican Republic and the U.S.A. and coffee rust spread from West Africa to South America (Kahn, 1989). Other important natural vectors are insects, animals and waterways. However, plants and plant parts including seeds are among the most important vehicles.

New or endemic pests and diseases?

It is important to distinguish between reports of new pests in a country and the actual interception of the pest. In many countries there is not sufficient knowledge of the pest situation. Systematic surveys are frequently lacking and reports are usually published only if a pest causes substantial yield losses. Information on pests which are endemic and inconspicuous is scarce. However, changed agricultural practice could promote epidemics, which may then be blamed on newly introduced pests. An example is the bacterial stripe disease, or black chaff, caused by Xanthomonas campestris pv. translucens in Syria. The symptoms are rather inconspicuous at a low disease incidence and severity. Symptoms on leaves may be confused with those of barley stripe mosaic virus, barley yellow stripe virus or the stripe disease in barley caused by Pyrenophora graminea. A black discoloration of the glumes in wheat or barley can also be caused by Alternaria spp., Cladosporium spp., or by Septoria nodorum. In Syria the disease was first observed in the late seventies in the areas of Raqqa and Hassakeh, and some scientists suspected that the pathogen had been introduced with seeds. However, the occurrence of the disease coincided with the implementation of extended irrigation in this area. Irrigation enhances infection of Xanthomonas, and to date the disease does not play any role in rainfed cereals in Syria. In this case the possibility of a newly introduced pathogen cannot be excluded, but is unlikely.

A clear affirmation is usually only possible when the pest is intercepted, i.e. isolated from imported seeds. This is especially true for most pests which have their own ways of distribution and many pathogens which may be transmitted by other means than seed (e.g. soil, wind, etc.). Even with exclusively seed-borne pathogens, such as Ustilago nuda or Pyrenophora graminea, infected plants in a field sown with imported seed could result from volunteer seed in the field originating from a crop in which the disease remained unnoticed, or from inadvertent seed mixtures.

Interception vs. transmission

Interceptions have to be looked at carefully. Pests introduced into new areas may not necessarily find the environmental conditions conducive to attack host plants, to multiply, or to survive and become endemic. An example is the introduction of the dwarf bunt fungus, Tilletia controversa, into warm agro-ecological zones. The fungus requires temperatures of 1-5°C for germination and grows best if the plants are under snow cover for extended periods of time. Although long persistence of the spores in the soil, compared to those of Tilletia caries and T. foetida, increases the risk of a disease outbreak in a year with favourable conditions, it seems quite unlikely that the pathogen could become established in a country such as Egypt. The same is true for many insects from warmer areas. They may multiply during the summer in a temperate climate but usually do not survive the winter. An example is the introduction of the larger grain borer, Prostephanus truncatus, to Iraq in 1969, which did not get established there (Laborius, 1988). However, pests may come in strains with different requirements for temperature and it is therefore possible that such a strain becomes established in an area previously not considered suitable for the survival of the pest.

If a pest is intercepted in a shipment from a particular country, the assumption is that the pest occurs in that country. However, pests may infest shipments en route and the common practice of re-forwarding seeds with a new phytosanitary certificate makes it difficult to pinpoint the country from which the seeds came originally. An example: between 1970 and 1980 the Indian quarantine service intercepted Tilletia indica in seeds received from, among other countries, Lebanon, Syria, Turkey, and Sweden (Lambat et al., 1983a). These countries are now frequently listed as host countries (e.g. Martin, 1989). An extensive survey of material from Syria (Diekmann, 1987), however, did not give any indication that this bunt fungus occurs in Syria, and the disease was not reported from any of the other countries mentioned. In the case of ICARDA it was confirmed that the seeds intercepted in India originated from a country where the disease occurs and were repacked and dispatched from Syria. The seeds from Lebanon, Turkey and Sweden were also dispatched from research institutes or genebanks and possibly did not originate from these countries either.

Examples of new pests in the West Asia and North Africa (WANA region)

With the caveat resulting from the paragraphs above some examples for pests that have spread to the WANA region from outside or inside the region from one country to another are given in Table 1.

Table 1. Some examples of pests newly introduced to countries of West Asia and North Africa.

Pest/pathogen	Introduced to	Reference
Xanthomonas campestris pv. translucens	Arab Republic of Yemen	Diekmann and Aalders du Bois, 1988
Ascochyta fabae	Egypt	Omar, 1986
Orobanche crenata	Ethiopia	Telaye and Saxena, 1986
Prostephanus truncatus	Iraq	Laborius, 1988

Although none of the pests have so far been reported to cause any serious crop losses in their new host countries, they have to be considered potentially important. It has to be expected that several years pass between the first report of a new pest and the development of an epidemic. The pests listed in Table 1 are all reported to be significant in countries where they are endemic. Duveiller (1989) summarized data on distribution of and yield losses caused by Xanthomonas campestris pv. translucens. About 40% losses were recorded in sprinkler irrigated wheat fields in Idaho, U.S.A., and 43% in CIMMYT tritical trials. In New Zealand, planting of faba bean seeds with a 12% infection with Ascochyta fabae resulted in a 44% yield loss (Gaunt and Liew, 1981). For Orobanche crenata on faba beans in Syria, the yield loss was estimated to reach between 5 and 100% (Sauerborn and Saxena, 1987). In Morocco, this parasitic weed caused catastrophic yield losses in the late seventies, and susceptible legumes are no longer planted in infested areas (Schlüter and Aber, 1980). Prostephanus truncatus, the larger grain borer, spread to Africa from Central America and the southern states of the U.S.A., where it is considered a minor pest. In Africa it has been reported to cause losses in stored maize of 30 to 35% (Makundi, 1987; McFarlane, 1988).

It remains to be seen whether the newly introduced pests will disappear again from their new host countries, as was the case with Prostephanus truncatus in Iraq, or will become endemic without causing severe yield losses, or will develop into pests of epidemic proportions. Besides prevailing cultivars and agricultural practices, climatic conditions are the determinant factors.

Quarantine efforts

Scientists and technologists working with seeds know that seed lots completely free from dangerous pests are impossible to find. This is the reason why plant quarantine organizations of different countries must have effective laboratory seed health testing facilities and well-managed post-entry control systems. Developing countries in general lack such facilities. Governmental authorities of different countries are becoming aware of the problems connected with indiscriminate movement of seed. In some developing countries such as India, Indonesia, Kenya, Malaysia, Nigeria, Philippines, South Korea and Thailand quarantine facilities have been strengthened.

With the improvement in seed health testing facilities several important pests such as Ustilago tritici, Urocystis tritici, Peronospora manshurica, Puccinia carthami, Fusarium nivale, Heterodera schachtii, H. goettingiana, Rhadinaphelenchus cocophilus, Oscinella frit, Acanthoscelides obtectus, Bruchophagus glycorrhizae, Ephestia elutella, Carpocapsa pomonella were intercepted in germplasm material at the National Bureau of Plant Genetic Resources of India in New Delhi (Lambat et al., 1983b). It has been mentioned by the authors that as far as possible infested material was salvaged using chemical and physical means.

The international agricultural centres have become more and more concerned about the dissemination of pests with seed of crop germplasm they distribute annually to cooperators, worldwide. Appropriate measures are taken to raise healthy crops in field, conduct thorough field inspection, collect seed from healthy plants/plots, clean the seed properly, subject seed for routine laboratory seed health testing and treat the seed by effective pesticides before distribution. Such facilities have been installed at the International Rice Research Institute (IRRI, Los Banos, Philippines), International

Crops Research Institute for the Semi-Arid Tropics (ICRISAT, Hyderabad, India), Centro Internacional de Mejoramiento de Maiz y Triego (CIMMYT, Mexico), International Center for Agricultural Research in the Dry Areas (ICARDA, Aleppo, Syria) and others.

While this article was under preparation, we wrote to some colleagues to gather the latest information on problems connected with importations. Information provided by a few is presented here. Professor Zheng Zhang, China, (Plant Pathologist of the China Animal and Plant Quarantine Headquarters, Beijing) wrote that Karnal bunt fungus of wheat, Tilletia indica, was intercepted in wheat germplasm received from CIMMYT, Mexico, in 1983 and all infected lots were destroyed. In 1990, another important bunt fungus of grasses, Tilletia fusca, also not known in China, was intercepted in seeds of Bromus and Festuca imported from the U.S.A.; all infested lots were destroyed. Dr J. Bucha of Mauritius (Plant Pathology Division, Ministry of Agriculture, Fisheries and Natural Resources, Réduit) informed us that a number of consignments of capsicum, sesame, maize, sugar beet and beans were destroyed because they were found infected with Ascochyta capsici, Alternaria sesami, Bipolaris maydis, Phoma betae and Macrophomina phaseolina, respectively. These interceptions were made during 1980 and 1984 by trained seed health testing analysts. Mrs Ruth Mathu of Kenya (Plant Quarantine Station, Muguga) told us that Ascoshyta blight of chickpea caused by Ascochyta rabiei was observed in a field at Katumani (KARI Station for Dryland Research) from seeds imported from India. The disease was eradicated but rules had to be changed. All chickpea seed consignments should now pass through the quarantine laboratory at Muguga. It is well known that attempts to introduce chickpea as a new crop to Canada were given up as Ascochyta blight was introduced with the seed (source unknown) and it was extremely destructive in yield trials in 1973 (Morall and McKenzie, 1974). Information received from Dr O.R. Reddy, India (Plant Quarantine Station, Bombay) on interceptions is summarized below:

- Downy mildew of soybean (Peronospora manshurica) from USA. Consignment destroyed, 1981.
- Pitch canker of slash pines (Fusarium moniliforme var. subglutinans) from Fiji, on 0.5% seeds in 1981. Consignment destroyed.
- Dwarf bunt of wheat (Tilletia controversa) and ergot sclerotia of Claviceps purpurea in bulk shipments from USA in 1988-89.
- Downy mildew of sunflower (Plasmopara halstedii) intercepted in 3 hybrids vars. imported from France in growing-on tests.
- Bacterial leaf spot (Pseudomonas syringae pv. maculicola) in 2 consignments of cabbage from USA, 1.5-2.0%.
- Tobacco mosaic virus (TMV) in two consignments of tomato seed imported from USA, 5-6% in growing-on tests. Also infection by bacterial stem canker was noted (Clavibacter michiganensis subsp. michiganensis).

Undoubtedly, it is difficult to measure the success of quarantine. Bartlett (1990) reported that the U.K. is free from the Colorado beetle (Leptinotarsa decemlineata) due to surveillance of imports and eradication campaigns that controlled a total of 143 out-breaks since 1877. Measures taken included a temporary ban on parsley imports from Italy in 1984.

Many of the Pacific Islands, including Australia and New Zealand, are still free from a number of potential pests (Dale and McKenzie, 1989).

At least temporarily successful attempts were the eradication campaigns for the Mediterranean fruit fly (Ceratitis capitata) in Florida and California which were carried out at great cost (Johnson and Schall, 1989). Islands are generally in a more advantageous position to eradicate pests, as shown by Dale and McKenzie (1989) with many examples, including that of stem nematode, coffee rust and citrus canker.

A case-history of Bipolaris maydis

Southern leaf blight disease of corn, caused by Bipolaris maydis, became well known in the world when Race T of the fungus attacked plants having Texas male-sterile cytoplasm in the U.S.A. leading to devastating losses. The fungus is seed-borne and it can be detected easily in the laboratory by the most common incubation test, the blotter method.

During routine seed health testing of about 6000 seed samples of different crops at the Danish Government Institute of Seed Pathology for Developing Countries by the blotter method, seed of black gram (Vigna mungo), cowpea, (Vigna unguiculata), green gram (Vigna radiata), lettuce (Lactuca sativa), paspalum (Paspalam scorbiculatum), sorghum (Sorghum vulgare) and watermelon (Citrullus lanatus) were found infected; percentage infection ranged from 0.3 to 2.8 (Fatima et al., 1974). The fungus did not produce disease on respective hosts, but all isolates produced mild to severe symptoms on standard differentials of maize. Five of the nine isolates tested were of Race T and four of Race O. Cultures of Race T were isolated from seeds of black gram, cowpea, paspalum and sorghum. The authors discussed ways in which seeds of these crops could have become contaminated or infected. Whatever may have been the means, infected seed of these non-host crops can act as potential source of bringing inoculum of B. maydis to both uninfected and infected areas. This case-history is important for plant quarantine workers as they should test not only maize seeds with care but also seeds of other crops for B. maydis. Already, there are two reports where sorghum seeds from Thailand and the U.S.A., carrying B. maydis were intercepted in India (Race T) (Ram Nath et al., 1973) and New Zealand (Scott, 1971).

Often during routine seed health testing an analyst finds a fungal species in seeds of a variety of crops, e.g. Fusarium moniliforme. This fungus can be found in seeds of cereals, pulses, vegetables, flowers and forest trees. What type of decision should a quarantine worker take in this situation? Morphologically, the fungus looks the same on seeds of different hosts. There may be, however, differences in their pathogenic behaviour. The decision always rests with the quarantine officer. The problem becomes still more difficult and complex when it comes to seeds of wild types. We do not know of any published examples where scientists have experimentally showed differences between isolates obtained from seeds of cultivated and wild types. This is an extremely interesting area open to research scientists. It is possible that seeds of wild type may have been responsible for introducing new "forms" in some countries as "silent introductions".

Concluding remarks

Whenever plant material is moved from one location to another, there is always a risk of pest movement along with it. In many cases this risk may be negligible, in others the expected benefit from introducing new material may outweigh a possible risk. Cost of running an efficient quarantine service has to be taken into consideration, too. In order to reduce such cost, it is important to concentrate on important "high-risk" pests. We agree to the justification for high-risk pests as proposed by Kahn (1989):

An exotic organism may be considered as high risk (or of quarantine significance) to an importing country

- if the organism does not occur in the importing country, and
  is not widely distributed in the ecological range of its hosts in that country
  or
  has a life and host/pest interaction indicating that the organisms is capable of causing damage under favourable host, environment, and pest population or pathogen inoculum circumstances, which occur in the importing country
  or
- if the organism occurs in an importing country, and the organism
  is not widely distributed in the ecological range of its hosts in that country
  or
  is under a national domestic suppression, containment, or eradication programme
  or
  has exotic strains of quarantine importance that do not occur in the country
  or
- if the organism is a common pest already established in the importing country, but government or industry agreement requires that commercial growers use pathogen-tested nursery stock, and the government has taken steps to insure that imported stocks meet domestic standards.

Proper risk analysis of pests in a quarantine station can only be done if the station is provided with a good infra-structure of facilities and a well trained staff. It is our hope that the workshop will discuss these fundamental inputs in detail.

We further hope that the delegates will give serious thought to the following two suggestions:

- to consider agro-ecological zones rather than countries. It is quite likely that the pest population does not change at borders. Only in special cases, freedom from a pest in one country justifies extensive routine testing of seeds coming from a neighbouring country.

- to list as pests of quarantine significance only those which are not already present in the country and could be of potential economic importance. Ubiquitous and saprophytic fungi, such as Cladosporium sp., should not be considered in quarantine.

References

Bartlett, P.W. 1990. Interception of Colorado beetle in England and Wales, 1983-1987. EPPO Bulletin 20: 215-219.

Dale, P.S. and McKenzie, E.H.C. 1989. Plant Quarantine - A Pacific Island Viewpoint. pp 169-193. In: Plant Protection and Quarantine. Vol. III. (ed. R.P. Kahn). CRC Press, Florida, U.S.A.

Diekmann, M. 1987. Monitoring the presence of Karnal bunt (*Tilletia indica) in germplasm exchange at ICARDA. Phytopath. Medit.* ***26: 59-60.***

Diekmann, M. and Aalders du Bois, E. 1988. Statement about the presence of a bacterial disease(Xanthomonas translucens) on barley in ARA Station, Dhamar. *Arab and Near East Plant Protection Newsletter* 6: 23.

Duveiller, E. 1989. Research on *Xanthomonas translucens* of wheat and triticale at CIMMYT. *EPPO Bulletin* **19:** 97-103.

Fatima, R., Mathur, S.B. and Neergaard, Paul 1974. Importance of *Drechslera maydis* on seeds of crops other than maize. *Seed Sci. & Technol.* **2:** 371-383.

Gaunt, R.E. and Liew, R.S.S. 1981. Control strategies for *Ascochyta fabae* in New Zealand field and broad bean crops. *Seed Science and Technology* **9:** 707-715.

Johnson, R.L. and Schall, R.A. 1989. Early Detection of New Pests. pp. 105-116. In: *Plant Protection and Quarantine*. Vol. III. (ed. R.P. Kahn). CRC Press, Boca Raton, Florida, U.S.A.

Kahn, R.P. 1989. *Plant Protection and Quarantine*. Vol. I. CRC Press, Boca Raton, Florida, U.S.A.

Laborius, G.A. 1988. The Larger Grain Borer Prostephanus truncatus (Horn) (Coleoptera: Bostrichidae) - A Quarantine Pest. In: *Movement of Pests and Control Strategies* (eds. K.G. Singh *et al.*). ASEAN Plant Quarantine Center and Training Institute.

Lambat, A.K., Ram Nath, Mukewar, P.M., Majumdar, A., Indra Rani, Kaur, P., Varshney, J.L., Agarwal, P.C., Khetarpal, R.K. and Usha Dev 1983a. International spread of Karnal bunt of wheat. *Phytopath. Medit.* **22:** 213-214.

Lambat, A.K., Wadhi, S.R. and Sanwal, K.C. 1983b. Seed health testing at the National Bureau of Plant Genetic Resources, India. *Seed Science and Technology* **11:** 1249-1257.

Kakundi, R.H. 1987. The spread of *Prostephanus truncatus* (Horn) in Africa and measures for ots control. *FAO Plant Protection Bulletin* **35:** 121-126.

Martin, P.M.D. 1989. The Pathogens of Wheat. pp. 1-22. In: *Plant Protection and Quarantine* (ed. R.P. Kahn) Vol. II. CRC Press, Boca Raton, Florida, U.S.A.

McFarlane, J.A. 1988. Pest management strategies for *Prostephanus truncatus* (Horn) (Coleoptera, Bostrichidae) as a pest of stored maize grain: present status and prospects. *Tropical Pest Management* **34:** 121-132.

Morall, R.A.A. and McKenzie, D.L. 1974. A note on the inadvertent introduction to North America of *Ascochyta rabiei*, a destructive pathogen of chickpea. *Pl. Dis. Reptr.* **58:** 342-345.

Neergaard, P. 1979. *Seed Pathology.* Vol. I and II. The Mcmillan Press. London.

Omar, S.A.M. 1986. Occurrence of Ascochyta blight of faba bean in Egypt. *FABIS* **15:** 48-49.

Ram Nath, Lambat, A.K., Payak, M.M., Lilaramani, J. and Rani, I. 1973. Interception of race T of *Helminthosporium maydis* on sorghum seed. *Curr. Sci.* **42:** 872-874.

Sauerborn, J. and Saxena, M.C. 1987. Weed survey of faba bean fields in Syria with special reference to *Orobanche* spp. and *Cuscuta* spp. *FABIS* **17:** 38-40.

Schlüter, K. and Aber, M. 1980. Chemical control of *Orobanche crenata* in commercial culture of broad beans in Morocco. *Zeitschrift für Pflanzenkrankheiten und Pflanzenschutz* **87:** 433-438.

Scott, D.J. 1971. The importance to New Zealand of seed-borne infection of *Helminthosporium maydis. Pl. Dis. Reptr.* **55:** 966-968.

Telaye, A. and Saxena, M.C. 1986. Orobanche in faba bean fields, in Ethiopia. *FABIS* **14:** 33.

International and Regional Cooperation in Plant Quarantine

The International Plant Protection Convention

MAHMOUD M. TAHER* and EDWIN FELIU**

*FAO Regional Plant Protection Officer for the Near East; ** Plant Quarantine Officer, Plant Production & Protection Division, FAO, Rome

Introduction

Wherever crops are grown, they have always been inter-related with their pests, i.e. insects, mites, fungi, nematodes, bacteria, viruses, mycoplasmas, weeds and others. Such pests can be of major or minor importance. However, if they exist in one locality they may not necessarily be present in another. Man's first known trade is believed to have been based on plants and plant products for human nutrition and basic agro-industry. Transportation of plants and plant produce in such trade then also naturally meant transportation of pests associated with them from one area to another. Some of the transported pests might not survive the new environmental conditions and hence are eliminated, whilst other minor pests might find themselves in a more suitable environment and become major pests. Fortunately, in general, up to the middle of the last century, movement of plants/plant products was limited to restricted areas, mainly due to non-existence of long-range means of transportation or problems connected with long-distance movement of such plants/plant products which could not withstand journeys of several weeks or months.

This whole picture has changed with the development of modern means of transport and travel, i.e. aeroplanes, steamships and refrigerated containers, which have made transport of plants and plant products quick and safe. The global need to introduce new crops, high-yielding varieties, and germplasm for improving plant quality, continues to grow. With such developments, the spread of new pests from one country to another, and from one continent or region to another, is increasing and, in many cases, has resulted in a wide range of problems of crop losses, leading to shortage of production, famine, and instability or collapse of industries based on plant products, and the application of expensive plant protection measures.

Pest introductions due to international movement of plants and plant products

The destruction of crops due to the movement of plant pests from one area to another is very well-documented. The following are a few examples of such records. In all these cases, man is the primary agent of movement and introduction of pests.

a) **Food grains**. The devastating Khapra beetle (*Trogoderma granarium*) is a good example of movement of pests in international trade. Its introduction was recorded in the USA, Italy and Zimbabwe in the 1940s and 1950s. This pest is spreading continuously in international trade. The larger grain borer (*Prostephanus truncatus*), a native pest of Central America and Mexico, was intercepted in Israel in 1962 and in Iraq in 1970 on maize consignments. This pest was introduced and has been well-established since the early 1980s in a number of East and West African countries where it causes severe damage. The wheat bunt (*Tilletia caries*) moved to the USA from Australia in 1854 through seed introductions. In addition, many weed species have been introduced and established in various regions of the world through cereal and other seed consignments.

b) **Vegetables and legumes**. The introduction of seed potatoes infected with the late blight pathogen (*Phytophthora infestans*) in the 1840s from Peru to Ireland and northern Europe is a well-documented case of destruction by exotic pests. This led to the total loss of potato crops resulting in famine and the death of millions of people. Other examples of diseases spread via the export and/or import of seed are: tomato bacterial canker (*Clavibacter michiganense* subsp. *michiganensis*) introduced into the UK from the USA in 1942; onion smut (*Urocystis cepuli*) from France to Switzerland in 1924; bacterial blight of soybean (*Pseudonomas syringae* pv. *glycinea*) from Sweden to Scotland in the 1940s; and bacterial blight of bean (*Xanthomonas campestris* pv. *phaseoli*) from the Netherlands to New Zealand in 1969.

c) **Fruit trees and vines**. Introduction of infected pome fruit planting material from the USA to Europe resulted in the spread and establishment of the fire blight pathogen (*Erwinia amylovora*) in many European countries, and its later spread to some North Africa, resulted in large-scale losses of pome fruits. The spread of the plum pox virus, known as Sharka, from Bulgaria to western Europe and to some north and south Mediterranean countries had a devastating effect on stone fruits. The San José insect (*Quadraspidiotus perniciosus*) is another pest which was introduced originally from China into the USA and then moved to Europe and South Africa, causing significant losses in fruit crops. Citrus tristeza virus, which is believed to have originated in China, has been moved via infected planting material to South Africa, Brazil, the USA, Spain and Israel, resulting in the destruction of tens of millions of citrus trees.

The movement of the Mediterranean fruit fly (*Ceratitis capitata*) through infested fruits and vegetables resulted in the spread of the insect, which has a wide range of fruit and vegetable hosts, to almost all continents of the world where it inflicts great crop losses requiring very expensive control and eradication programmes.

d) **Industrial crops**. Tobacco blue mould (*Peronospora tabacina*) was introduced from Australia and the USA to England, and then to Europe and North Africa, resulting in extensive damage to the tobacco industry. Also, the introduction of the coffee rust pathogen (*Hemileia vastatrix*) from Africa to Sri Lanka in 1857 led to almost total destruction of coffee fields. The disease later spread to many South and Central American countries. The cotton pink bollworm (*Pectinophora* *gossypiella*) was first recorded in India in 1842 and has spread from India to major cotton-growing areas of the world, inflicting important economic losses. Examples of diseases spread via the export/import of seed are beet rust (*Uromyces betae*) from Europe to Canada in 1943; flag smut of wheat (*Urocystis agropyri*) from Australia to Mexico in 1919; and posm disease of flax (*Septoria linicola*) from Canada to Australia.

In addition to the above examples of introduction and spread of plant pests, cases of interception of plant pests from one country/ region/continent to another continue to be reported.

The introduction, spread and establishment of pest problems through international trade, small consignments carried by individual travellers or exchanged for scientific purposes, and the consequence of such introduction expressed in terms of crop destruction and losses, entail substantial expenditure for eradication and control measures, that could amount to tens of millions of dollars and, in many instances, are not effective or efficient.

The concern of the international community since the 1880s to check pest movement was expressed by the establishment of the following conventions:

a) **International Convention against *Phylloxera vastatrix*.** The introduction of grapevine phylloxera into France through importations of American grapes in 1859 and its spread throughout Europe, resulting in destruction of vineyards and the collapse of viticultural industries, led to the development of a Convention to prevent further introductions and spread of this pest. The Convention was signed on 3 November 1881 by five countries, and was later adhered to by 12 countries. However, this Convention was disregarded after the successful use of resistant American rootstocks and the movement of grapevine cuttings without roots as a means of preventing spread of *Phylloxera.*

b) **1929 International Convention of the Protection of Plants**. The need for international cooperation for preventing the introduction of pests through the importation of living plants or seeds has been reiterated since 1891 in international meetings, and an International Act to provide such cooperation was adopted during an International Conference of Phytopathology held in Rome on 4 March 1914. However, the First World War halted its development. It was not until 1929 that modifications to the above Act were brought up in the International Plant Protection Conference held in that year, where a final draft of the International Convention of the Protection of Plants was signed. Due to the fact that only 17 countries adhered to the Convention, its usefulness was limited.

The above agreements were a first indication of international awareness of the importance of quarantine as a safeguard to agricultural development and food security.

The need for effective, modern quarantine measures to match the development of modern methods of transportation and the increased need for improving quantity and quality of plant produce, which necessitates the movement of plants and their produce, emerged in the late 1940s. Governments recognized the usefulness of international cooperation in controlling pests of plants and plant products, and the prevention of their introduction and spread across national/international boundaries. With the desire to ensure close coordination of measures directed to these ends, the Sixth Conference of the Food and Agriculture Organization of the United Nations (FAO), held in November 1951, approved the International Plant Protection Convention (IPPC).

The 15 Articles of the Convention

Article I. Purpose and Responsibility. The purpose of the Convention is to ensure effective international cooperation to contain the introduction and spread of plant pests and disease, and for this, specific legislative, administrative and technical measures are required to be undertaken by governments which should also assume the responsibilities within their territory for all IPPC requirements.

Article II. Scope. The Convention makes particular reference to pests and diseases of importance to trade, and its provisions extend to storage places, containers, conveyances, packing material and accompanying media including soil. Definitions of plants and plant products were also provided.

Article III. Supplementary Agreements. Supplementary agreements applicable to specific regions, pests and diseases and/or plants and plant products or means or methods of transport may be, on government recommendation, proposed to FAO and come into force after acceptance in accordance with FAO rules and procedures.

Article IV. National Organizations for Plant Protection. To control pests and diseases of economic importance, the Convention makes it a pre-requesite that contracting governments establish national plant protection services to: undertake inspection of the phytosanitary status of growing plants as well as plant products; survey pests and diseases and report on outbreaks; inspect plants and plant products moving in international trade with the objective of preventing the dissemination of pests and diseases; disinfect or disinfest such plants or plant products and their ambient as necessary; issue phytosanitary certificates in relation to conditions of the consignment; create awareness on plant pests and diseases and means of their control; and undertake research and investigation for plant protection.

Article V. Phytosanitary Certificates. Contracting governments should arrange for issuance of phytosanitary certificates, according to the model annexed to the Convention, after inspection by qualified personnel and in accordance with the regulations of the importing government.

Article VI. Requirements in relation to Imports. In order to prevent the introduction of pests and diseases into their territory, contracting governments may set provisions to regulate a) entry of plants and plant products including restrictions on importation, prohibition of importation of particular plants/plant products, detention of particular consignments, treatment and destruction or rejection of entry of infected/infested consignments, and list pests subject to prohibition of entry and/or treatment; b) minimize interference with international trade through undertaking above as necessary, publish and communicate to FAO and concerned Regional Plant Protection Organizations (RPPOs) such prohibition/restriction with reason(s), communicate promptly intercepted problems to exporting country, minimize certification requirements, and establish adequate safeguards for importations for scientific purposes; and c) that FAO disseminate received information on importation restrictions, requirements, prohibitions and regulations as received from contracting governments to other contracting governments and Regional Plant Protection Organizations.

Article VII. International Cooperation. Contracting governments shall agree to cooperate with FAO in establishing a global reporting service on pests and diseases and, in this context, provide reports on the existence, spread and outbreaks of quarantine-significant pests and means for their control. They should also participate in campaigns against pests requiring international action.

Article VIII. Regional Plant Protection Organizations. In order to ensure coordination and promotion of regional and inter-regional plant protection activities, the Convention makes provisions for the establishment of regional plant protection organizations.

Article IX. Settlement of Disputes. If conflicts arise between contracting governments on quarantine matters related to import/export of plants and plant products within the provisions of the Convention, FAO will appoint a Committee to consider the question under dispute. The Committee's recommendations will serve as a basis for consideration of settlement of the dispute.

Articles X to XV. These articles make provision for substitution of prior agreements; territorial application; ratification and adherence; amendments; entry into force; and denunciations.

The approved International Plant Protection Convention came into force on 3 April 1952 under ratification by three signatory governments and, by 1 May 1952, 37 FAO Member Nations had signed the Convention. FAO is the depository of this Convention and administers its provisions. Currently 96 FAO Member Countries are signatories of the IPPC (Annex I).

Undoubtedly the provisions of the International Plant Protection Convention have made a valuable worldwide contribution to plant protection activities in general and to arresting the spread of plant pests in particular. Signatories and, equally, many non-signatories of the Convention, in particular the developing countries, have established Plant Protection Organizations; undertaken surveys, studies and research programmes on pests of economic importance; developed control strategies and quarantine services; enacted legislation and regulations; trained personnel; installed required facilities for inspection, detection and treatment of pest problems associated with import and export of plants and plant products; exchanged information between concerned parties directly and through RPPOs and FAO; identified pests of quarantine significance and monitored their spread and established safeguards for their introduction; and established Regional Plant Protection Organizations in almost all regions of the world (Annex II), thereby initiating coordination and harmonization of plant protection activities among Member Countries on the one hand, and cooperating and coordinating plant protection activities on a global scale on the other. FAO, for its part, as depository and administrator of the Convention, has established the Plant Protection Bulletin as a global reporting service on legislation, and incidence and spread of plant diseases and pests and means for their control. It coordinates the activities of Regional Plant Protection Organizations and assists Member Countries in establishing functional plant protection and quarantine services, setting up plant quarantine data bases, and formulating guidelines for germplasm transfer.

Revised IPPC

After 20 years from the date that the IPPC came into force, and in the light of plant quarantine developments that have occurred during that period of time and the requirements of international trade, *Ad hoc* consultations were held in 1973 to re-examine the provisions of the IPPC and consider recommendations for amendments. In November 1976, a Government Consultation was convened in Rome by the Director-General of FAO to consider proposed amendments. These were accepted and submitted to the 20th Session of the FAO Conference in November 1979, when the revised text was approved.

The main amendments introduced into the Convention are:

a) Replacement of the terms "contracting governments" and "adhering governments" with "contracting party(ies)" and "adhering state(s)", "pests and diseases", "pests or diseases" with the term "pest(s)". The definition of the terms "pest" and "quarantine pest" was also adopted.

b) Paragraph 3 of Article II (formerly paragraph 2) was modified to start with "where appropriate" and to include "any material capable of harbouring and/or spreading plant pests particularly those involved in international transportation".

c) In Article VII on Requirements in relation to imports, in 1., a new paragraph (e) was added, to read "list pests whose introduction is prohibited or restricted because they are of potential economic importance to the country concerned". In the same article in various paragraphs in 2, notification of decisions on regulations, prohibitions, requirements etc. should be notified to FAO and "any regional plant protection organization of which the contracting party is a member and all other contracting parties directly concerned". In addition, point 4 was added and reads "FAO shall disseminate information received on information restrictions, requirements, prohibitions and regulations (as specified in paragraph 2(b), (c) and (d) of this Article) at frequent intervals to all contracting parties and regional plant protection organizations".

d) The phrase, "RPPOs shall gather and disseminate information" was added to the second paragraph of Article VIII on Regional Plant Protection Organizations.

e) In the new Phytosanitary Certificate the wording of the testimony has been significantly changed to include the proviso that plants or plant products described have been inspected according to appropriate procedures and are considered to be free from quarantine pests, and practically free from other injurious pests etc., etc., instead of the statement that "... were thoroughly examined and were found, to the best of the inspector's knowledge, to be substantially free from injurious diseases and pests ..."
In addition to the model phytosanitary certificate, the revised IPPC introduced a new model phytosanitary certificate for re-export. The re-export phytosanitary certificate is required, together with the original phytosanitary certificate when the consignment has entered into temporary storage, is repacked, or is treated.

The revised text of IPPC entered into force on 4 April 1991, after ratification by two-thirds of the signatory parties.

GATT and IPPC

At present the General Agreement on Tariffs and Trade (GATT) has recently shown interest in ensuring that phytosanitary measures are not used as trade barriers. Due to GATT's lack of expertise in this field, and based on provisions of Article VI of the Convention, GATT is calling on IPPC to advise on such matters and, in particular, to define more closely the "burden of proof" or the "burden of justification" when quarantine action is taken. In this respect, GATT hopes that the IPPC could set standards and/or guidelines that could potentially be used in a large number of countries and which could serve as a reference point in the settlement of disputes. GATT is currently working on the framework of an agreement which requires FAO and RPPO inputs. In the light of the above, since 1989 FAO initiated regular Technical Consultations among Regional Plant Protection Organizations, with the objectives of a) promoting harmonization of quarantine measures, particularly phytosanitary principles, procedures and pest risk assessment in order that phytosanitary conditions are not used as unjustifiable trading restrictions, and b) amelioration of information exchange and notification. As a result of these consultations, the required guidelines are being prepared and coordination of plant protection activities, including information exchange and notification among RPPOs, is underway.

Conclusions

Throughout recent history, the movement of plant pests in international trade and germplasm exchange have contributed towards crop losses and increasing costs of crop protection all over the world. Such a situation has rightly created awareness of the importance of international cooperation to prevent the introduction and spread of exotic plant pests. As a result, a number of international conventions have been established to control the movement of pest(s) of quarantine significance, the earliest of which dates back to 1881. The currently implemented International Plant Protection Convention has provided a basis for the development of plant quarantine principles and procedures, and has established a role for Regional Plant Protection Organizations. Doubtless, full and proper implementation of the IPPC will pave the way for cooperation and coordination of global efforts to prevent the introduction and spread of quarantine significant pests.

The IPPC should always be considered as a dynamic instrument requiring constant revision to cope with the continued development of quarantine requirements.

The GATT Uruguay Round of negotiations has emerged at the right time to give us a gentle reminder that proper implementation of the IPPC is essential in the modern world.

Annex I. Contracting parties to the International Plant protection Convention

Algeria, Argentina, Australia, Austria, Bahrain, Bangladesh, Barbados, Belgium, Belize, Bolivia, Brazil, Cambodia, Canada, Cape Verde, Chile, Colombia, Costa Rica, Cuba, Czechoslavakia, Denmark, Dominican Republic, Ecuador, Egypt, El Salvador, Ethiopia, Finland, France, German Federal Republic, Ghana, Greece, Grenada, Guatemala, Guyana, Haiti, Hungary, India, Indonesia, Iran, Iraq, Ireland, Israel, Italy, Jamaica, Japan, Jordan, Kenya, Republic of Korea, Laos, Lebanon, Liberia, Libyan Arab Jamahirirya, Luxembourg, Mali, Malawi, Malta, Mauritius, Mexico, Morocco, Netherlands, New Zealand, Nicaragua, Niger, Norway, Oman, Pakistan, Panama, Papua New Guinea, Paraguay, Peru, Philippines, Portugal, Rumania, Saint Christopher & Nevis, Senegal, Sierra Leone, Solomon Islands, South Africa, Soviet Union, Spain, Sri Lanka, Sudan, Suriname, Sweden, Thailand, Togo, Trinidad & Tobago, Turkey, Tunisia, United Kingdom, United States of America, Uruguay, Venezuela, Yemen Arab Republic, Yugoslavia and Zambia.

Annex II. Regional Plant Protection organizations

Asia and Pacific Plant Protection Commission (APPPC) (Established in 1956)

Member Governments Australia, Bangladesh, Burma, Democratic Kampuchea, Fiji, France, India, Indonesia, Korea (Rep. of), Laos, Malaysia, Nepal, New Zealand, Pakistan, Papua New Guinea, Philippines, Portugal, Samoa, Solomon Islands, Sri Lanka, Thailand, Tonga, United Kingdom, Viet Nam

Caribbean Plant Protection Commission (CPPC) (Established in 1967)

Member Governments	Barbados, Colombia, Costa Rica, Cuba, Dominica, Dominican Republic, France, Grenada, Guyana, Haiti, Jamaica, Mexico, Netherlands, Nicaragua, Panama, St. Christopher & Nevis, St. Lucia, Suriname, Trinidad & Tobago, United Kingdom, United States of America, Venezuela

Comité Regional de Sanidad Vegetal para el Cono Sur (COSAVE) (Established in March 1989)

Member Governments	Argentina, Brazil, Chile, Paraguay, Uruguay

European and Mediterranean Plant Protection Organization (EPPO) (Established in 1950)

Member Governments	Algeria, Austria, Belgium, Bulgaria, Cyprus, Czechoslovakia, Denmark, Federal Republic of Germany, Finland, France, German Democratic Republic, Greece, Guernsey, Hungary, Ireland, Israel, Italy, Jersey, Luxembourg, Malta, Morocco, Netherlands, Norway, Poland, Portugal, Romania, Spain, Sweden, Switzerland, Tunisia, Turkey, Union of Soviet Socialist Republics, United Kingdom, Yugoslavia

Interafrican Phytosanitary Council (IAPSC) (Established in 1956)

Member Governments	Algeria, Angola, Arab Republic of Egypt, Benin, Botswana, Burkina Faso, Burundi, Cameroon, Central African Republic, Chad, Comoros, Congo, Côte d'Ivoire, Djibouti, Equatorial Guinea, Ethiopia, Gabon, Gambia, Ghana, Guinea, Guinea Bissau, Kenya, Lesotho, Liberia, Libya, Madagascar, Malawi, Mali, Mauritania, Mauritius, Morocco, Mozambique, Niger, Nigeria, Rwanda, Sao Tome & Principe, Senegal, Seychelles, Sierra Leone, Somalia, Sudan, Swaziland, Tanzania, Togo, Tunisia, Uganda, Zaire, Zambia

Junta del Acuerdo de Cartagena (JUNAC) (Established in 1969)

Member Governments Bolivia, Colombia, Ecuador, Peru, Venezuela

North American Plant Protection Organization (Established in 1976)

Member Governments Canada, Mexico, United States of America

Organismo Internacional Regional de Sanidad Agropecuaria (OIRSA) (Established in 1955)

Member Governments Costa Rica, El Salvador, Guatemala, Honduras, Mexico, Nicaragua, Panama

South Pacific Commission (SPC) (Established in 1947)

Member Governments American Samoa, Australia, Commonwealth of Northern Mariana Islands, Cook Islands, Federated States of Micronesia, Fiji, France, French Polynesia, Guam, Kiribati, Marshall Islands, Nauru, New Caledonia, New Zealand, Niue, Norfolk Island, Palau, Papua New Guinea, Pitcairn Islands, Solomon Islands, Tokelau, Tonga, Tuvalu, United Kingdom, United States of America, Vanuatu, Wallis & Futuna Islands, Western Samoa

Literature consulted

FAO. 1951. *International Plant Protection Convention*. Rome.

FAO. 1974. *Revised International Plant Protection Convention*. Rome.

FAO. 1983. *Plant Quarantine Guide*. Rome.

FAO. 1990. *Report of the Second Technical Consultation between Regional Plant Protection Organizations*. May 21-25, 1990. Rome.

Hewitt, W.B. & Chiarappa L. (eds.). 1977. *Plant Health and Quarantine in International Transfer of Genetic Resources*. CRC Press, Cleveland, Ohio.

Howe, R.W. 1972. Insects attacking seeds during storage. pp. 247-300. In: *Seed Biology*. Vol. III. (ed. Kozlowski).

Karpati, J.G. 1983. Plant quarantine on a global basis. *Seed Science & Technology* **11**: 1145-1157.

Ling, L. 1953. International Plant Protection Convention: its history, objectives and present status. *FAO Plant Protection Bulletin* **1**(5): 65-68.

Neergaard, P. 1979. *Seed Pathology*. Vol. I & II. The Macmillan Press, London and Basingstoke. 1191 pp.

Quarantine for Seed: Status, Requirements and Implications

EDWIN FELIU* and MAHMOUD M. TAHER**

*Plant Quarantine Officer, Plant Production & Protection Division, FAO, Rome; **FAO Regional Plant Protection Officer for the Near East

Introduction

The international exchange of germplasm and other propagative plant materials, including seeds, is needed in order to sustain valuable research and breeding programmes for increased crop production. However, it is well known that seeds may serve as pathways for the introduction and spread of plant pests. Reportedly, there are more than 1,000 host-pathogen combinations of seed-borne diseases caused by fungi, bacteria, nematodes, insects, and virus and virus-like organisms. In the case of seeds, many of these pests are seed-transmitted and have moved internationally, causing devastating damages to major crops. Some examples of these are:

- Rice blast, caused by *Pyricularia oryzae*, which was introduced into Africa with seed from Asia in 1971,
- *Ascochyta fabae*, introduced into Australia with broad bean from the UK in 1977,
- *Tilletia caries* (bunt of wheat), introduced from Australia into the USA in 1854,
- *Xanthomonas campestris* pv. *oryzae* (bacterial blight of rice), introduced into South America and the Caribbean with infected seed in the 1970s,
- *Xanthomonas campestris* pv. *manihotis* (cassava bacterial blight) has been proven seed-borne and its introduction into Africa is believed to have been through infected seed, and
- *Peronospora tabacina* (tobacco blue mould) was introduced into England in 1958 and spread to all tobacco growing areas of Europe, parts of North Africa and the Near East.

The aforementioned are only a few examples of quarantine pests that have moved across national boundaries and have become established in new areas. The list is very extensive and continually growing as the international exchange of seeds intensifies. Consignments of seeds vary from small packets to bulk shipments. For instance, the International Crops Research Institute for the Semi-Arid Tropics (ICRISAT) reportedly sent more than 4,000,000 seed samples internationally from 1974 to 1986. Other international research centres also send vast amounts of seeds around the world. Similarly, the International Rice Research Institute (IRRI) distributes approximately 150,000 rice seed samples yearly. For the quarantine organizations in many developing countries, the handling of these considerable amounts of seeds may present an operational problem. When tested, many seeds have shown varying degrees of infection with viruses, fungi and other pathogens. These pathogens are difficult to detect using ordinary inspection methods.

Unfortunately, most developed and developing countries do not conduct seed health tests on a routine basis, relying mainly on visual examination. As these pests often do not reveal their presence at the time of importation, they may go undetected and, thus, gain entry and become established.

In view of this substantial pest introduction risk, most countries react by either prohibiting the entry of "high-risk" categories or by requiring growing under post-entry quarantine for a specified period of time to observe the introduced materials under secluded conditions. Needless to say, both of these actions impose considerable restraints on the international exchange of seeds. Although most countries allow the importation of small amounts of prohibited plant materials for research under strict safeguards, and the type of seeds requiring post-entry are relatively limited, the movement of seeds is substantially impeded in these cases, resulting in an adverse impact to useful research and breeding programmes. Even in cases of less restricted categories, quarantine services have been blamed for undue delays in clearing germplasm which has resulted in its inevitable loss. Quarantine actions reportedly have sometimes been too drastic with plant materials found infested or infected with a plant pest which could have been eliminated and the material freed of it. The decisions in the latter cases have often been destruction of the plant materials, causing irreparable loss to importers.

On the other hand, quarantine services may not have another feasible alternative to assuming a conservative attitude when faced with a substantial pest risk situation which is beyond their expertise, facilities or capabilities to resolve. In these cases, adopting a conservative stand becomes a sounder decision than substantially endangering quarantine security. Under these circumstances, it seems reasonable to assume an attitude that duly considers the two genuine interests at play: the need to exchange germplasm and the recognized right of countries to protect themselves against pest introduction. If adequate safeguards are observed when handling germplasm, including seeds, the risk associated with such movement can be substantially minimized in order to provide quarantine security.

The nature and extent of such safeguards will vary according to the particular pest-host situation involved. For instance, for certain "very high risk" materials there are no known safeguards that can provide quarantine security and, hence, their importation may only be exceptionally allowed under very strict safeguards, in small amounts. The latter, however, should only be the case for very limited categories. For most other seeds, requirements may be so relaxed that certain flower and vegetable seeds may not even require import permits or phytosanitary certificates. Other "low risk" categories may only require a visual or microscopic examination, while certain consignments may have to undergo specific testing or treatment before release. A limited number of seeds may require growing under post-entry or to pass through an intermediate quarantine station before importation. The sound application of safeguards would ensure a reasonable level of quarantine safety while facilitating the exchange of planting materials. However, in order to apply appropriate safeguards, national quarantine services must be fully operational, that is, they must be strengthened to the point where they have trained staff and facilities for proper inspection, testing and treatment of seeds. As most national quarantine services in developing countries are grossly deficient in one or more of the above, it is not possible for them to implement the necessary safeguards. At present, the only recourse for these countries is to either exclude "high risk" materials, to require intermediate quarantine growing or to obtain materials only from reliable sources certified as being free of certain pests, such as viruses, by the exporting country. It is only through the strengthening of quarantine services to a fully operational level that these safeguards may be properly applied.

Elements of plant quarantine

The regulatory framework

Acts, rules and regulations are enacted by countries in order to restrict the movement of plant and plant products, their carriers, means of conveyance, packing materials and soil since they may serve as pathways for pest introduction or spread. Assuming that an appropriate pest risk assessment has been conducted, plant commodities are placed in unrestricted, restricted and prohibited categories. The quarantine safeguards vary from a visual inspection to treatment or complete exclusion (prohibition) for the above categories. Import permits specifying the necessary conditions of entry provide an opportunity for a country to assess the risk associated with a proposed importation and decide on its entry status.

Most countries have prepared lists of pests of quarantine significance (quarantine objects). These are usually pests not present or not widely distributed in the country which pose a substantial threat to its major crops. Likewise, most Regional Plant Protection Organizations (RPPOs) have also prepared such lists for their respective regions. The quarantine pest lists provide a useful reference point for exporting countries and constitute a valuable component of the quarantine regulatory framework. The lack of such lists leaves the decision as to the phytosanitary condition of the consignment to the exporting country. In order to be meaningful, however, pest lists must be continuously reviewed in the light of new information obtained through pest surveys or reports on the occurrence or outbreak of pests. Likewise, rules and regulations need to be periodically assessed to ensure their effectiveness. Prohibitions and restrictions must be fully justified, and areas of limited pest prevalence should be considered for possible determination of "pest-free" areas when supported by appropriately conducted surveys and other controls.

Facilities for inspection, detection and treatment

The effectiveness of quarantine safeguards depends to a large extent on the availability of adequate inspection, detection, treatment and disposal facilities. These include inspection equipment and materials such as microscopes, magnifying lenses, identification keys, pest collections and other reference materials. Also required are laboratory detection capabilities such as seed health testing, serology, indexing, X-rays and other facilities. Treatment facilities should include fumigation chambers, hot water tanks, chemical dipping tanks and dry heat ovens. Disposal equipment for potentially infected or infested seeds (autoclaves, incinerators, etc.) is also needed.

Some seeds may also require growing under post-entry quarantine requiring the construction of specialized facilities such as glasshouses, screenhouses and laboratories.

Unless these facilities are either in the quarantine service or available to them through reliable institutions within the country, it would not be possible to apply the necessary safeguards.

The use of facilities belonging to research centres, universities and other scientific institutions, could be a valuable resourse for the quarantine services of many developing countries which are unable to obtain these specialized facilities so that adequate quarantine control may be exercised.

Staffing

In order to fully implement quarantine safeguards, quarantine services need specially trained staff capable of detecting and eliminating cryptic quarantine pests such as bacteria, fungi, insects, nematodes, virus and virus-like organisms. In addition, if post-entry quarantine facilities are established, glasshouse horticulturists and other technical staff may be needed to ensure optimal conditions for plant growth.

Application of available methodologies

During various meetings held between Regional Plant Protection Organizations and FAO, considerable concern was expressed on the safe movement of germplasm. The need to develop protocols for such movement was indicated.

FAO and the International Board of Plant Genetic Resourses (IBPGR) jointly launched a programme to develop crop-specific guidelines for the Safe Transfer of Germplasm. These protocols have been established through meetings of recognized experts for the concerned crops. Protocols for bananas, cacao, root crops, legumes, citrus and cassava have been developed. Similar guidelines are planned for grapevine, vanilla, stone fruits and other crops.

The guidelines provide specific indications on a step-by-step basis on the appropriate handling of germplasm, from the time that the seed is harvested or collected to the time of the release of germplasm. They also provide specific information on the plant pests associated with the plant material and on available intermediate quarantine stations. The guidelines for legumes cover extensively the handling of seeds.

Although the guidelines are for voluntary adoption, it is expected that they will become internationally acceptable methodologies for the safe transfer of seeds and other propagative materials. These guidelines need to be uniformly applied when moving germplasm so as to ensure its safe and expeditious transfer.

Public awareness and support

The effectiveness of the aforementioned quarantine safeguards may only be secured through the implementation of a public awareness programme at national, regional and global levels.

Unless researchers, breeders and others engaged in the exchange of germplasm are made aware of the importance of safely moving germplasm, pests will continue to move through this pathway, as their cooperation is vital to the endeavour. International cooperation is vital to the success of this endeavour, particularly with the implementation of the International Plant Protection Convention (IPPC) of 1951. Some international research centres have appointed quarantine officers to ensure that appropriate safeguards are taken when moving germplasm. This should be emulated by other centres involved in the exchange of germplasm.

Information on plant quarantine

As explained previously, regulatory decisions must be preceded by a sound pest risk assessment. However, in order to conduct this analysis, the quarantine organization must have access to timely and reliable information on quarantine pests. This should comprise the geographical distribution and biological characteristics of the concerned pests. Information on host associations, pest descriptions, symptoms and behaviour is necessary in order to properly assess pest potential in the light of available staffing, facilities and other safeguards.

Unfortunately many countries do not have access to such vital information and, consequently, their pest risk analysis capabilities are considerably limited. In addition, information on pest outbreak occurrences and interceptions is not readily communicated, as required by the relevant provisions of the IPPC.

Implications of the increased exchange of seeds

In view of the enhanced exchange of germplasm and other planting materials, and the inability of quarantine organizations to improve accordingly, a global threat of pest dissemination exists. Urgent measures have to be implemented in order to avert the crisis. Plant quarantine still constitutes the safest and most economical way to prevent pest introduction and spread.

Logically, quarantine organizations need to be strengthened to a fully operational level. This means that they should be able to implement all the safeguards mentioned previously and to have access to updated information on plant pests.

Moreover, international cooperation needs to be fostered at regional and global levels. This is best done through the effective implementation of the provisions of the International Plant Protection Convention of 1951. In doing this, it should be kept in mind that the exchange of seeds and other germplasm by international research centres, breeders and all those involved in their international movement should be facilitated. This involvement must include close cooperation with national and regional quarantine organizations as well as with FAO and other international organizations concerned with plant protection. Only through these concerted efforts may germplasm and seeds be moved safely and expeditiously.

Literature consulted

Berg, G.H. 1989. *La Cuarantena Vegetal: Teoría y Práctica.* OIRSA, 1989. 440 pp.

FAO. 1951. *International Plant Protection Convention.* Rome.

FAO. 1974. *Revised International Plant Protection Convention.* Rome.

FAO/DANIDA Seminar on Quarantine for Seed for Developing Countries of Africa and Asia. New Delhi, November 19 - December 2, 1980. Proceedings.

FAO/IBPGR. 1988 and 1989. Guidelines for the Safe Exchange of Germplasm. Guidelines for Cocoa and Legumes.

Hewitt W.J. and Chiarappa, L. (eds.). 1979. *Plant Health and Quarantine in International Transfer of Genetic Resources.* pp. 290-306. CRC Press.

International Consultation on a System for Safe and Efficient Movement of Germplasm. June 15-17, 1982. Cali, Colombia. Proceedings.

Joshi, N.C. and Joshi, R.C. 1989. Role of Plant Quarantine in Crop Pest Management in India, a review. *Agric. Rev.* **10**(4): 167-182.

Kahn, R.P. 1979. A Concept of Pest Risk Analysis. *EPPO Bulletin* **9**(1): 119-130.

Kahn, R.P. 1989. *Plant Protection and Quarantine.* Vol. I. CRC Press. 226 pp.

Kahn, R.P. (ed.) 1989. *Plant Protection and Quarantine.* Vol. II. CRC Press. 265 pp.

Kahn, R.P. (ed.) 1989. *Plant Protection and Quarantine.* Vol. III. CRC Press. 215 pp.

Karpati, J. 1983. *Plant Quarantine on a Global Basis.* FAO, Rome. 30 pp.

Morschel, J.R. 1971. *Introduction to Plant Quarantine.* Australian Publishing Service, Canberra.

Neergard, P. 1972. International and national cooperation in seed health testing and certification. In: *Proceedings of the International Seed Testing Association* **37**: 117-138.

Neergard, P. 1980. A review on quarantine for seed. National Academy of Sciences, India. *Golden Jubilee Commemoration Volume*: 495-530.

Phatak, H.C. 1981. *International Transfer of Germplasm.* FAO Publication No. RAPA 50.

Plucknett, D.L. and Smith, N.J.H. 1988. Plant quarantine and the international transfer of germplasm. *World Bank Study Paper* No. 25: 1-21.

Varma, B.K. and Ravi, U. 1984. Plant quarantine facilities developed at ICRISAT for export of germplasm. *Plant Protection Bulletin* **36**(2/3): 37-43.

EPPO's Experience of Quarantine for Seed

I.M. SMITH

European and Mediterranean Plant Protection Organization, 1 Rue le Nôtre, 75016 Paris, France

Introduction

EPPO, the European Plant Protection Organization, is the regional plant protection organization for Europe and the Mediterranean area, in the sense of the International Plant Protection Convention. Nearly all the governments of Europe are members, together with many Mediterranean countries, including those of the Maghreb. Although EPPO concerns itself in all the activities of Plant Protection Services in general, its main and original aim has been to prevent the spread and introduction of pests - in other words plant quarantine. In this respect, the EPPO Working Party on Phytosanitary Regulations has been meeting since EPPO's foundation 40 years ago to develop EPPO's particular plant quarantine strategy.

The basic element of this strategy, in line with the 1979 revision of the IPPC, is to take phytosanitary measures against quarantine pests, as required by the revised Convention. Each country thus has to define which are its quarantine pests, and EPPO recommends that a list of quarantine pests should be included in their regulations. This should not include all possible quarantine pests for a country, since they are numerous, but must include those which merit special attention. For each of these, the country should specify the measures to be taken by the exporting country in order to be able to export consignments of the plants and plant products concerned. These required measures should be made known to all potential exporters (multilaterally).

EPPO has gone further in advising its Member Governments which are the recommended quarantine pests for the region - by establishing the so-called A1 and A2 lists (Mathys and Smith, 1984). The A1 list covers quarantine pests which are not present in any EPPO country. They are considered as exotic pests, against which all countries should take measures, to ensure by solidarity that they are excluded from the whole region. The A2 list covers pests which have a restricted distribution in the EPPO region - so that they are absent from some countries, or else are "not widely distributed and being actively controlled" (in the words of the IPPC definition). Member Governments are invited to decide which pests of the A2 list they should cover in their regulations, but discouraged from adding any other pests of limited distribution in the region. EPPO has also made recommendations to its Member

Governments on the requirements to be made for these pests (specific quarantine requirements, or SQRs). These relate to the measures to be taken in exporting countries for specified commodities of plants and plant products liable to carry the pest. They may offer alternative, equivalent measures that exporting countries may take, or may specify different measures according to the pest status of the country concerned. Finally, EPPO has prepared data sheets on each of the quarantine pests of the A1 and A2 lists, and is also publishing "quarantine procedures", which are detailed instructions for performing treatments, tests, surveys etc.

Quarantine measures for seeds

EPPO's strategy does not make any special provision for seeds. Certain quarantine pests are liable to be carried on seeds, and in that case the specific quarantine requirements will provide for measures to be applied to seeds. It happens to be the case that many pests which are seed-borne are not or hardly transmitted by other means - and conversely, pests which are carried on rooted plants, or on fruits and vegetables, are not seed-borne. As a result, there is a fairly distinct category of seed-borne quarantine pests, set apart by the fact that their SQRs relate only to seeds. It also happens that a number of quarantine pests can be carried by any rooted plant, which means that all such plants are covered by specific quarantine requirements, and in particular require phytosanitary certificates. In contrast, seed-borne pests are rather host specific, so that only seeds of the relevant host plants are covered by SQRs. As a result, phytosanitary certificates are needed only for those seeds. It should be noted that EPPO's strategy is in contrast to others adopted elsewhere in the world : EPPO recommends its Member Governments not to require phytosanitary certificates for most seeds.

Seed-borne quarantine pests

Table 1 gives a list of the EPPO A1 and A2 quarantine pests which are seed-borne, together with their host plants. This shows immediately that relatively few hosts are concerned by the current quarantine pests. Some of them (rice, cotton, soybean) concern rather few countries : overall, it may be said that the total number of seed-borne quarantine pests for EPPO is not very high, and that individual countries are concerned by even fewer. This is reflected in the fact that many countries regulate imported seeds under different legislations, and enforce the regulations through a service other than the plant protection service. Nevertheless, when quarantine pests are concerned, EPPO recommends that the plant protection service is the competent service to enforce phytosanitary regulations applied to seeds.

Requirements for seed-borne quarantine pests

The EPPO-recommended requirements for seed-borne pests are generally not very strignent (Table 1). They most commonly offer a choice between a field inspection of the seed crop or a seed test. In more detail, the requirements abbreviated in the

Table 1. Seed-borne quarantine pests on the EPPO A1 and A2 lists.

Pathogen	Host	A1/A2	Quarantine requirements*	Test**
Aphelenchoides besseyi	rice	A2	T	++
Barley stripe mosaic virus	barley	A2	FI or T	++
Clavibacter michiganensis subsp. *insidiosus*	lucerne	A2	FI and special	
subsp. *michiganensis*	tomato	A2	FI or Tr	++
Cochliobolus carbonum	maize	A2	FI or T	++
Curtobacterium flaccumfaciens pv. *flaccumfaciens*	bean	A2	AF or FI	+
Erwinia stewartii	maize	A2	FI	+
Glomerella gossypii	cotton	A2	FI or T or Tr	++
Listronotus bonariensis	grass & cereals	A1	PC	
Mycosphaerella linicola	flax	A2	FI	
Phaeoisariopsis griseola	bean	A2	FI or T	++
Phialophora gregata	soybean	A1	Clean	
Phytophthora megasperma f. sp. *glycinea*	soybean	A2	Clean	
Potato viruses (non-potato European)		A1	P	++
Pseudomonas syringae pv. *pisi*	pea	A2	AF or FI	+
Stenocarpella macrospora and *S. maydis*	maize	A2	FI or T	++
Tilletia controversa	wheat	A2	FI	
Tilletia indica	wheat	A1	FI and T	++
Xanthomonas campestris pv. *corylina*	hazelnut	A2	AF or FI	
pv. *oryzae*	rice	A1	FI and T	+
pv. *oryzicola*	rice	A1	FI and T	+
pv. *phaseoli*	bean	A2	FI or T	+
pv. *vesicatoria*	tomato	A2	(FI+Tr) or T	++

* AF = area freedom, FI = field inspection, P = prohibition, PC = phytosanitary certificate only, T = test, Tr = treatment

** Published (++) or in preparation (+)

Table can be set out as follows :

- Area freedom - the consignment must come from an area where the pest does not occur.
- Field inspection - the consignment must come from a crop which was found free from the pest during the growing season.
- Prohibition - importation prohibited, except with a specially negotiated import permit.

- Phytosanitary certificate - will be needed in any case, to justify that requirements are met. In some cases, it is alone sufficient (i.e. visual inspection only).
- Test - the consignment must have been tested by an EPPO-recommended procedure.
- Treated - the consignment must have been treated by an EPPO-recommended procedure (usually HCl extraction).

Only in the case of some A1 pests are the requirements more stringent: countries have an option to prohibit, and if they do accept, require two conditions to be satisfied, A and B rather than A or B.

EPPO quarantine procedures

When a test has to be used (a frequent case for seeds), EPPO tries to develop a recommended test procedure (so-called EPPO quarantine procedure). In Table 1, the cases are shown where a procedure has been published or is in preparation. It may be noted that procedures are in preparation for several cases where testing is not given as a requirement. This reflects the very rapid current advances in methods for diagnosis and detection (Smith, 1990). If the new methods are considered suitable, the SQRs will be revised to allow for their use.

The EPPO Quarantine Procedures are not strict standards, but rather provide advice on how to conduct tests or to choose from alternative tests. It may be noted that the FAO/IBPGR germplasm guidelines provide a very valuable source of similar information.

Zero tolerance

The phytosanitary certificate requires freedom from quarantine pests, and this has usually been interpreted as a requirement for a zero tolerance. So long as test methods are relatively insensitive, this poses no problem - and has been operated in practice - a seed test being no better or worse than another requirement to be fulfilled before the certificate can be signed. However, new technology is making tests extremely sensitive, and this raises the question whether previously undetectable contamination is now a basis for rejecting consignments. Smith (1990) has argued that, if a known and agreed quarantine procedure is followed, the real phytosanitary guarantee is better now than it used to be. What is needed is minimum risk, not an un-realizable zero tolerance.

EPPO's approach to seed-borne quarantine pests

As outlined above, EPPO does not recommend its Member Governments to make requirements for most seeds, and recommends fairly simple requirements when it does. EPPO sees no necessity for tests for "quality pests", these pests which affect the quality of the consignment (and crop grown from it) but which are not quarantine pests, and do not pose a risk for the importing countries.

Even in the case of quarantine pests, EPPO perceives the risk as often rather less than for other quarantine pests. Many seed-borne pests (but not all) do not persist in the soil - so, even if an infested crop is grown, the pest will probably disappear subsequently. Provided seed crops are not contaminated, an importing country can usually tolerate an occasional outbreak coming from an imported seed-lot. For example, Italy has had several outbreaks of *Erwinia stewartii* over the years, but there is no evidence that the disease has established (and there is no vector for it in Europe). Similarly, Europe is developing soybean production and is beginning to produce its own seed. It is the seed-production system which must be kept clean, which is why requirements for imported soybean seed can be kept fairly minimal. Of course, seed exporting countries face a further problem, that their international markets can be affected even by a single reported outbreak of a pest, which could lead to new requirements and increased costs. Thus, Europe certainly should avoid any outbreak of *Tilletia indica*, though it is doubtful whether this would cause very serious loss to European crops as such.

Finally, EPPO does not seek to introduce new barriers to the seed trade, if this can be avoided. For example, EPPO recently assessed *Xanthomonas campestris* pv. *translucens* as a quarantine pest. The question is still under review, but was perceived as problematic, since no agreement could be reached on requirements to be made for imported cereal seed. It was suggested that countries which were concerned should use internal measures to test such seed, imported as germplasm, rather than propose requirements to exporting countries.

Re-export

Another problem for EPPO countries, as seed importers and exporters, lies in re-exporting seed to countries which do require phytosanitary certificates for seed. In principle, the country of origin should be known, and information should be available from it to show that requirements of the country of final destination can be satisfied. This is often very difficult to operate and practice, and EPPO is currently studying how to handle this seed re-export question.

References

Mathys, G. and Smith, I.M. 1984. Regional and global plant quarantine strategies with special references to developments within EPPO. *OEPP/EPPO Bulletin* **14**(2): 83-96.

Smith, I.M. 1990. New techniques for detection and identification of seed-borne pathogens in relation to international seed exchange and the zero tolerance concept. *Seed Science and Technology* **18**: 461-465.

Seed-borne Diseases of Economic Significance

Seed-borne Diseases of Wheat and Barley

OMAR F. MAMLUK

Cereal Improvement Program, International Center for Agricultural Research in the Dry Areas (ICARDA), P.O.Box 5466, Aleppo, Syria.

Introduction

Exchange of seed, especially among breeders is essential for the maximum utilization of genetic variability in plants. Uncontrolled exchange of germplasm may, however, lead to introduction of serious exotic plant pathogens into new areas. Seed-borne pathogens are sometimes the cause of disease outbreaks, since small amounts of seed-borne inoculum can be of great epidemiological significance. Even low seed transmission of pathogens may be important if seed is sown in uninfested soil.

Seed-borne diseases

Excluding some foliar diseases, especially rusts and powdery mildews, the great majority of wheat and barley diseases are seed-borne.

Table 1 gives an overview of the major seed-borne diseases of wheat and barley in the countries of West Asia and North Africa, the WANA region. The references cited are by no mean complete, but represent those mentioned in documents available at the time of preparation of this manuscript. Diseases encountered by the author in the WANA region (Mamluk, 1983-1991) are also included. As a rule, references that cite all previous references from a country, as in Mamluk *et al.* (1984 and 1990) has been used. The number of countries from which these seed-borne diseases have been reported indicates the occurrence of a disease in the country and its frequency of occurrence in the region, but does not necessarily reflect the importance of that particular disease. Relevant information on the frequency of occurrence of these diseases and their distribution are from the author.

The most frequently reported smut diseases of wheat and barley in the WANA region are loose smuts, *Ustilago tritici*, and *Ustilago nuda*, respectively; covered smut of barley, *U. hordei*; common bunt of wheat, *Tilletia foetida* and *T caries*; and flag smut of wheat, *Urocystis agropyri*. Flag smut has also been reported on barley in Syria (Azmeh and Kousaji, 1982). Dwarf bunt, *T. controversa*, which is limited in distribution to areas with snow cover during winter, has been reported from Turkey,

Iran and Iraq. Semi-loose smut of barley, *U. nigra*, has been reported from Lebanon and Jordan.

Karnal bunt, *T. indica*, has not been found in Syria (Diekmann, 1987; Mamluk *et al.*, 1990) as previously reported by Lambat *et al.* (1983) and Zillinsky (1983).

From the Helminthosporia group, net blotch, *Pyrenophora teres*, followed by barley leaf stripe *P. graminea*, are the most frequently reported diseases of barley in the region. Tan spot, *P. trichostoma*, has been reported from Turkey, Iran, Algeria, Tunisia and Morocco.

Septoria tritici blotch (*Mycosphaerella graminicola*) is the most widely distributed Septoria disease in the region. Septoria nodorum blotch *(Leptosphaeria nodorum)* though recorded in four countries, pathogenicity tests confirming its identification have been reported only from Syria. The pathogen has been found at only two sites of Syria, but with very low incidence (Mamluk *et al.*, 1990). Septoria blotch on barley, *Septoria passarini*, has been reported only from Syria (Mamluk *et al.*, 1983; Anonymous, 1984-1988). Scald, *Rhynchosporium secalis*, is one of the most common diseases of barley in WANA.

Bacterial leaf streak on wheat and barley, *Xanthomonas campestris* pv. *translucens* has been reported from Pakistan, Turkey and Syria. The disease was found recently in the desert areas of Fazzan in Libya (Mamluk 1983-1991). Other bacterial diseases reported are the basal glume rot on wheat, *Pseudomonas syringae* pv. *artofaciens*, and the bacterial leaf blight on barley, *P. syringae* pv. *syringae*, from Syria.

Barley stripe mosaic virus has been reported from Syria, Lebanon and Jordan (Makkouk and Jarikji, 1983), but studies at the Center by Makkouk indicate the presence of this virus disease in most countries of the WANA region (K.M. Makkouk, personal communication).

Wheat seed gall nematode, *Anguina tritici*, has been reported from West Asian countries, Pakistan, Turkey, Iraq, Syria and Jordan. The first record of this nematode on barley was from Iraq (Stephan, 1988).

Diseases of wheat and barley grown in the newly developed irrigation areas, such as the dry areas on the Euphrates in Syria and the desert areas of Fazzan in Lybia are mainly seed-borne. These areas are isolated from the old traditional ones; however, through dam irrigation and pivot systems man has produced excellent environmental conditions for disease development in these areas. Many of the cultivars introduced into these areas are vulnerable to diseases and with the help of seed transmitted pathogens, diseases like loose smuts, covered smut, barley leaf stripe, scald, net blotch and bacterial leaf streak have become endemic to these areas (Mamluk, 1983-1991).

Table 1. Major seed-borne diseases of wheat and barley in countries of West Asia and North Africa, the WANA region.

Disease	Country
Smuts	
1. Common bunt (*Tilletia foetida* & *T. caries*)	Pakistan, Egypt, Iran, Morocco Turkey, Syria, Lebanon, Jordan
2. Dwarf bunt (*Tilletia controversa*)	Turkey, Iran, Iraq
3. Loose smut * (*Ustilago tritici*)	Pakistan, Egypt, Iran, Libya, Turkey, Morocco, Syria, Lebanon, Jordan
4. Flag smut (*Urocystis agropyri*)	Pakistan, Egypt, Turkey, Tunisia, Syria, Jordan
5. Covered smut (*Ustilago hordei*)	Pakistan, Egypt, Turkey, Libya, Iran, Algeria, Iraq, Morocco, Syria, Jordan
6. Loose smut * (*Ustilago nuda*)	Pakistan, Egypt, Turkey, Libya, Iran, Morocco, Iraq, Syria, Lebanon, Jordan
7. Semi-loose smut (*Ustilago nigra*)	Lebanon, Jordan
Helminthosporia	
8. Barley leaf stripe * (*Pyrenophora graminea*)	Turkey, Libya, Iran, Algeria, Syria, Morocco, Lebanon, Jordan
9. Net blotch (*Pyrenophora teres*)	Pakistan, Egypt, Turkey, Libya, Syria, Tunisia, Jordan, Algeria, Cyprus, Morocco
10. Tan spot (*Pyrenophora trichostoma*)	Turkey, Tunisia, Iran, Morocco
Septoria	
11. Septoria tritici blotch (*Mycosphaerella graminicola*)	Pakistan, Tunisia, Syria, Algeria, Lebanon, Morocco
12. Septoria nodorum blotch (*Leptosphaeria nodorum*)	Syria, Morocco, Lebanon, Jordan
13. Septoria blotch (*Septoria passarini*)	Syria
Scald	
14. *Rhynchosporium secalis*	Turkey, Libya, Iran, Tunisia, Iraq, Algeria, Syria, Lebanon, Jordan

.... cont.

Disease	Country
Bacteria	
15. Bacterial leaf streak (*Xanthomonas campestris* pv. *translucens*)	Pakistan, Libya, Turkey, Syria
16. Basal glume rot (*Pseudomonas syringae* pv. *atrofaciens*)	Syria
17. Bacterial leaf blight (*Pseudomonas syringae* pv. *syringae*)	Syria
Viruses	
18. Barley stripe mosaic virus * (BSMV)	Syria, Lebanon, Jordan
Nematodes	
19. Seed gall nematode * (*Anguina tritici*)	Pakistan, Turkey, Iraq, Syria, Jordan

* exclusively seed-borne

Root rots caused by *Fusarium* and *Cochliobolus* spp. are not included in this list due to their complexity and their involvement in other diseases, such as leaf and seedling blight and black point.

The wheat flag smut, loose smut and septoria blotch were found on wild relatives of wheat, *Aegilops* spp., in Syria. However, the role of these hosts in the survival and build-up of the pathogen and in the epidemiology of these diseases as well as the possible seed transmission has not been clarified (Mamluk *et al.*, 1990).

Some of the above mentioned diseases of wheat and barley are also soil-borne or can survive on crop residues and thus have another mode of transmission. The control of such diseases is complicated. Diseases that are exclusively seed-borne are the loose smuts of wheat and barley, barley leaf stripe, barley stripe mosaic virus and the seed gall nematode.

Smut diseases

Smuts are the most serious group of seed-borne diseases in the countries of WANA. Even when yield losses are minor, smut contamination of the grain is often sufficient to reduce quality and cause marketing problems. In smuts, particularly loose smuts, disease incidence (% infected heads) is equated directly to yield loss.

We do not have a clear picture of the losses caused by smuts in the WANA region. In Table 2 an attempt has been made to estimate losses and give information related to crop loss such as damage, infested areas, or disease incidence caused by smuts.

Losses of 1 to 7% are common. High smut incidence has been reported from many countries. In some regions of Turkey, dwarf bunt causes losses as high as 80%. The use of ineffective chemical control against the soil-borne inoculum might have contributed to these high losses. The second most important cause of losses is loose smut of wheat, possibly due to the use of untreated seed or the use of non-systemic chemicals.

Appropriate chemical seed treatment is a very effective measure to control most smuts; however, Hoffmann (1982) estimated that less than 40% of the seed sown in WANA is treated. Cost of chemical seed treatment and distribution of treated seed are the major constraints. In low-input agriculture with barley as sole crop, the farmer still does not use certified and treated seed for every planting season, nor does he have the equipment to treat the seed.

The widely used fungicides for seed treatment to control smut diseases of wheat and barley in WANA are the compounds carboxin (Vitavax), in combination with captan (Vitavax-300) or with thiram (Vitavax-200); copper oxyquinoleate (several trade names); and terbutrazole (Raxil 2DS; Folicur). Till now there is no report of resistance of smut pathogens to these compounds from the WANA region.

Compared to other diseases studies on genetic variability in smut pathogens are relatively few in the WANA region. Genetic variability in common bunt pathogens has been studied in Pakistan (Mirza and Khan, 1983), Turkey (Parlak, 1981), Egypt (Kamel *et al.*, 1979) and Syria; on loose smut in Turkey (Parlak, 1981); on flag smut in Pakistan (Khan *et al.*, 1984) and Egypt (El-Khadem *et al.*, 1980). This rarity of research is probably due to the assumption that smuts are easily controlled by chemical seed-treatment. Thus a low research priority has been allocated for investigating their genetic variability.

Sources of resistance and resistant cultivars have been identified for common bunt (Abdel-Hak and Ghobrial, 1975a; Parlak, 1981; Mamluk and Nachit, 1988); loose smut of wheat (Bassiouni *et al.,* 1988); covered smut (Abdel-Hak *et al.,* 1975); and dwarf bunt (Parlak, 1981). It is assumed that "..modern-high yielding cultivars, might have less bunt resistance than the land varieties used previously" (Hoffman, 1982). Many of the old cultivars possessed excellent resistance to smut diseases. Cultivar Tosson, which was grown in Egypt, is resistant to common bunt (Abdel-Hak and Ghobrial, 1975a) and loose smut of wheat (Bassiouni, 1988). In Syria, the once widely grown cultivar, Senatore Cappelli (known in North Africa as Jenah Khatifa) possesses excellent resistance to common bunt (Mamluk and Nachit, 1988). Also the famous durum wheat land race from Syria and Jordan "Haurani" has reasonable resistance to common bunt.

Table 2. Estimates of the area infested and incidence of smut diseases and the consequent loss and damage in some countries of the WANA region

Disease	Region/ Country	Infested area/ Disease incidence/ Loss/Damage	Reference
Common bunt	WANA	5-7% loss	Hoffmann, 1982
	Turkey	10% of fields infested; damage reaches 60-90% in some fields.	Parlak, 1981
		15-20% loss	Yuksel *et al.*,1980*
	Syria	50% of fields surveyed infested with incidence up to 60% in a field.	Mamluk *et al.*,1990
		5-7% loss	Mamluk *et al.*,1989
	Pakistan	25-50% disease incidence.	Mirza and Khan, 1983*
Dwarf bunt	Turkey	Infested area covers 2500 ha.	Parlak, 1981
		Up to 80% loss in some regions.	Yuksel *et al.*, 1980*
Loose smut (wheat)	Turkey	32% of fields infested; damage 16%; 35,000 tons loss.	Parlak, 1981
	Egypt	90% of fields with cv Sakha 61 infested with average incidence 0.01-0.1%	Bassiouni *et al.*, 1988
Flag smut	Turkey	little losses	Parlak, 1981
	Egypt	1% loss	Abdel-Hak & Ghobrial, 1975b
Common bunt and loose smut (wheat)	Syria	20% loss	Mulder, 1958
Covered smut	Egypt	1-2% loss	El-Said *et al.*, 1982
Barley smuts	Jordan	5-6% loss	Mamluk, 1984

* Reference citing other researchers

Almost all common bunt resistance now available in ICARDA's improved durum wheat cultivars has been derived from Senatore Cappelli (Mamluk and Nachit, 1988). Work needs to be continued to find new sources of resistance.

In a new development, partial infection of the spike is considered as a partial resistance to common bunt and studies to confirm this have been initiated at ICARDA.

Concluding remarks

1. Seed-borne diseases of wheat and barley are gaining more importance, probably due to:
   a) the availability of other sources of inoculum, soil-borne inoculum, which is not affected by any chemical seed treatment or the possible survival of the pathogen on crop debris and alternative hosts,
   b) ineffective chemical seed treatment, using old chemicals whose shelf life has expired or using non-systemic fungicides for pathogens which are located inside the seed,
   c) lack of seed treatment, and
   d) possible resistance of smut pathogens to chemicals.

2. Mechanized harvesting and the decrease of acreage planted with bunt resistant cultivars in some countries of the WANA region.

3. The tendency of mono-culturing wheat and barley without effective seed treatment or using susceptible cultivars is likely to increase incidence of seed-borne bunts and smuts, or may promote the establishment and build-up of soil-borne inoculum.

4. Most diseases in isolated agricultural areas, e.g. the newly developed irrigated schemes, are seed-borne and there are indications that these diseases were introduced by contaminated/infected seed.

5. Due to the current infrastructure and/or financial constraint, the use of chemical seed treatment or certified seed on large scale in many countries of the WANA is limited.

   The use of resistant cultivars for the control of smut diseases remains the most feasible and practical measure.

References

Abdel-Hak, T. and Ghobrial, E. 1975a. Studies on the control of bunt of wheat *Tilletia foetida* Waller. In: *Agri. Res. Rev.* **2:** 45-52.

Abdel-Hak, T. and Ghobrial, E. 1975b. Studies on the control of flag smut of wheat, *Urocystis tritici* Koern. In: *Agri.Res. Rev.* **2:** 35-43.

Abdel-Hak, T.M., Ghobrial, E. and Sabet, T. 1975. Studies on the control of covered smut of barley, *Ustilago hordei* (Pers.) Lagerh. In: *Agri. Res. Rev.* **2:** 53-60.

Anonymous 1984-1988. Cereal pathology. In: *Collaborative Research and Training Program.* ICARDA/DASR, Aleppo/Damascus, Syria.

Azmeh, M.F. and Kousaji, T. 1982. *Contribution to Survey of Plant Diseases in Syria, Unreported Fungal Diseases*. 22nd Science Week, Board of High Education, Ministry of High Education. Nov. 1982. Damascus, Syria. 16 pp.

Bassiouni, A.A., Abou El-Naga, S., Gouda, A. and Youssef, W.A. 1988. Sources of resistance to loose smut of wheat caused by *Ustilago tritici* (Pers.) Rostrup. In: *Proceedings of the 7th European and Mediterranean Cereal Rusts Conference*. Sept. 5-9, 1988. Vienna, Austria.

Bayya'a, B., Al-Ahmed, A. and Bellar, M. 1978. First list of unreported fungal diseases in districts Hama, Aleppo and Lattakia in Syria. *Pl. Prot. News* No. **3-4:** 1-10. Syrian Society for Plant Protection, University of Aleppo, Aleppo, Syria.

Diekmann, M. 1987. Monitoring the presence of Karnal bunt (*Tilletia indica*) in germplasm exchange at ICARDA. *Phytopath. Medit.* **26:** 59-60.

El-Khadem, M., Omar, R.A., Kamel, A.H. and Abou El-Naga, S.A. 1980. Physiologic and pathogenic races of *Urocystis agropyri*. *Phytopath. Z.* **98:** 203-209.

El-Said, H.M., Ghobrial, E., Ismail, H.A. and Rizk, R.A. 1982. Inheritance of reaction to covered smut disease of barley. pp. 19-30. In: *Proceedings, Egypt. Bot. Soc.* Mansoura, Egypt.

Hoffmann, J.A. 1982. Bunt of wheat. *Plant Disease* **66:** 979-987.

Kamel, A.H., Omar, R.A., El-Khadem, M. and Abou El-Naga, S.A. 1979. Physiologic and pathogenic races of *Tilletia foetida*. pp. 853-863. In: *Proceedings of the 3rd Egypt. Phytopath. Congress*.

Khan, M.A., Mirza, M.S., Hamid, S.J. and Khokhar, C.K. 1984. Physiologic races in flag smut of wheats. *Pakistan J. Agric. Res.* **5:** 172-174.

Lambat, A.K., Ram Nath, Mukewar, P.M., Majumdar, A., Rani, Indra, Kaur, P. Varshney, J.L., Agrawal, P.C., Khetarpal, R.K. and Dev, Usha 1983. International spread of Karnal bunt of wheat. *Phytopath. Medit.* **22:** 213-214.

Makkouk, K.M and Jarikji, O.A. 1983. Detection of sap-transmissible viruses infecting cereals in Jordan, Lebanon and Syria. *Journal of Plant Diseases and Protection* **90:** 12-17.

Mamluk, O.F. 1983-1991. Observations on wheat and barley diseases encountered in North Africa and West Asia, the WANA region. (unpublished data).

Mamluk, O.F. 1984. Review of smut diseases in Jordan. Pages 192-198. In: *Proceedings, Barley Diseases and Associated Breeding Methodology Workshop*. USAID-MSU/ICARDA/CIMMYT, April 20-23, 1981, Rabat, Morocco. Montana State University, Bozeman, Montana, USA.

Mamluk, O.F., Abu Gharbieh, W.I., Shaw, C.G., Al-Musa, A. and Al- Banna L. 1984. *A Check List of Plant Diseases in Jordan*. Publ. Univ. of Jordan, Amman. 107 pp.

Mamluk, O.F., Al Ahmed, M. and Makki, M.A. 1990. Current status of wheat diseases in Syria. *Phytopath. Medit.* **29**: 143-150.

Mamluk, O.F. and Nachit, M. 1988. Performance and reaction of some durum wheat genotypes against different isolates of common bunt (*Tilletia foetida* and *T. caries*). In: *Abstracts of papers, 5th International Congress of Plant Pathology*. August 20-27, 1988. Kyoto, Japan.

Mamluk, O.F., Bellar, M. and El-Naimi, M. 1983. *A List of Diseases of Forage Crops encounteredri Lebanon, Syria, Jordan, Tunisia, and Morocco (1981-1983)*. ICARDA, Aleppo, Syria. 6pp.

Mamluk, O.F., Haware, M.P., Makkouk, K.M. and Hanounik, S.B. 1989. Occurrence, losses and control of important cereal and food legume diseases in West Asia and North Africa. pp. 131-140. In: *Proceedings, 22nd International Symposium on Tropical Agriculture Research*. August 25-27, 1988. Kyoto, Japan.

Mirza, M.S. and Khan, M.A. 1983. A new race L-2, of *Tilletia foetida* from Pakistan. *Pakistan J. Agric. Res.* 4: 37-40.

Mulder, D. 1958. Plant diseases of economic importance in northern region, United Arab Republic. *FAO Plant Prot. Bull.* **7:** 1-5.

Parlak, Y. 1981. Seed-borne pathogens on wheat (particularly smuts). *EPPO Bull.* **11:** 83-86.

Stephan, Z.A. 1988. Plant parasitic nematodes on cereals and legumes in Iraq. pp. 155-159. In: *Proceedings of the Workshop on Nematodes Parasitic to Cereals and Legumes in Temperate Semi-arid Regions*. March 1-5, 1987. Larnaca, Cyprus.

Yuksel, H., Guncan, A. and Doken, M.T. 1980. The distribution and damage of bunts (*Tilletia* spp.) and wheat gall nematode [*Anguina tritici* (Steinbuch) Chitwood] on wheat in the eastern part of Anatolia. *J. Turkish Phytopath.* **9:** 77-88.

Zillinsky, F.J., 1983. *Common Diseases of Small Grain Cereals: a Guide to Identification*. International Maize and Wheat Improvement Center (CIMMYT), Mexico. 141 pp.

Diseases of Faba Bean and Lentil

S.B. HANOUNIK* and B. BAYAA**

*ICARDA, P.O. Box 5466, Aleppo, Syria; **Faculty of Agriculture, Aleppo University, Aleppo, Syria

Introduction

Faba bean (*Vicia faba*) and lentil (*Lens culinaris*) are attacked by a wide range of pathogens. Of the 147 species of pathogens reported on faba bean and lentil, only a few are seed-borne. The importance of these pathogens as a major constraint in faba bean, and to a lesser extent, in lentil production, has become increasingly evident during the past decade. In faba bean the wide prevalence of some pathogens has dictated fundamental changes in the cultivation and the trade of the crop. Some air-borne pathogens, for example *Botrytis fabae*, have forced farmers to give up faba bean cultivation in certain humid zones, whereas other seed-borne pathogens, such as *Ascochyta fabae* and *Ditylenchus dipsaci*, have caused the enforcement of new quarantine regulations which halted seed exports from infested areas.

In lentil, *Fusarium oxysporum* f.sp. *lentis*, *Uromyces fabae* and *Ascochyta lentis*, are the most important pathogens which can devastate the crop in certain areas and years. Most commercial cultivars grown by farmers today are susceptible, and breeding for resistance to diseases affecting faba bean and lentil has long been hampered by the lack of useful sources of resistance. This paper reviews the distribution and economic importance of major pathogens affecting faba bean and lentil, advances made in their control and discusses approaches to reduce their dissemination through seed.

Faba bean

Chocolate spot

Distribution and importance. Chocolate spot, caused by *Botrytis fabae*, is the most important disease of faba bean. It occurs almost anywhere faba bean is grown and has been reported from various parts of Europe, the Middle East, North Africa and Canada (Guant, 1983). Severe chocolate spot epidemics have occurred in Syria, Tunisia and England. In Syria, up to 75% loss has been reported from areas with extended periods of wet and cool weather conditions (Hanounik, 1979).

Epidemiology. The sexual stage of *B. fabae* has not been identified. The sclerotial and mycelial stages of the pathogen, in infected crop debris, are the most important forms of survival (Harrison, 1979). Conidia obtained from one-year-old refrigerated sclerotia of *B. fabae* were more virulent than those obtained from naturally infected faba bean leaves (Hanounik and Hawtin, 1982). *B. fabae* has been isolated from faba bean seed but no correlation was found between seed-borne inoculum and disease incidence in the field (Sode and Jorgensen, 1974). Harrison (1978) could not detect *B. fabae* in infected seed after nine months of storage in paper bags in the laboratory. Although occurrence of *B. fabae* in faba bean is common, it may not be important in the survival of the pathogen and the fungus may be eliminated after one year of seed storage (Harrison, 1978).

Severity of chocolate spot has been found to be greater in faba bean plots planted in December than those planted in February, and seven-week-old plants showed more severe symptoms compared to two- week-old plants under artificial inoculation in the field (Hanounik and Hawtin, 1982). In general, plant organs are less susceptible to *B. fabae* at the 10% podding than at the 100% podding stage, but leaf tissue is more susceptible than either stem or pod tissues at both podding stages. The disease becomes destructive during extended cool and wet weather conditions (Sundheim, 1973).

Control. The most practical method to control chocolate spot is to use resistant cultivars. Considerable progress in the area of breeding chocolate spot-resistant cultivars has recently been made as a result of the identification of durable resistance in the faba bean lines, BPL710 and 1179 (Hanounik and Maliha, 1986). Resistance from BPL710 and 1179 has been incorporated into well adapted genetic stocks (Robertson and Hanounik, 1987). In Egypt new faba bean cultivars with resistance from BPL1179 have recently been released. New sources of resistance to chocolate spot have been identified (Hanounik and Robertson, 1987).

Low seeding rate (Ingram and Hebblethwaite, 1976), delayed planting (Hanounik and Hawtin, 1982), elimination of infected plant debris after harvest (Harrison, 1979), and rotating faba bean with cereals play an important role in reducing disease severity.

Fungicides may be used to protect germplasm or early faba bean plantings that are aimed at high market price. In Syria, the use of vinclozoline (Ronilan 50WP) as a foliar spray, once every two weeks (total eight applications), controlled the disease and increased yield by 58% in treated plots compared to untreated plots (Hanounik, 1981).

Recently, Hanounik (1988) showed that the cultural filtrates of *Penicillium citrinum* and *P. cyclopium* suppressed spore germination of *B. fabae* and reduced disease development equally well, as did the widely used fungicide vinclozoline.

Ascochyta blight

Distribution and importance. Blight caused by *Ascochyta fabae* is one of the most important seed-borne diseases of faba bean. It has been reported from several locations in Europe, Asia, USSR, Australia, North Africa, Canada and Argentina (Guant, 1983). An epidemic was reported from the coastal area near Lattakia in Syria in 1976 (Hanounik, 1979). In New Zealand, a yield reduction of 44% was caused by *A. fabae* when seed with 12% infection was used for planting (Guant *et al.*, 1978).

Epidemiology. No sexual stage has been reported for *A. fabae*. Information on the survival of the pathogen is fragmentary. Dodd (1971) showed that *A. fabae* can survive for up to four months in infected debris buried in soil under field conditions. Recently, it is shown that *A. fabae* can survive longer in infected seeds in the laboratory, compared to infected buried or unburied stems in the field. The fungus was recovered from 95% of the seed after 12 months of storage and from only 5% of the seed after 27 months of storage. The pathogen remained viable in buried and unburied stems for only six and nine months, respectively. Hewett (1973) reported that 0-15% of infected faba bean seeds produced infected seedlings in the field. The causal relationship between seed infection and disease appearance in the field was reported by other workers (Guant and Liew, 1981).

Little is known about the effects of environmental factors on disease development, but extended cool and wet weather conditions are considered favourable (Sundheim, 1973).

Control. The use of clean seed is the most economic and efficient method of control (Guant and Liew, 1981). In the UK, the adoption of the 1969 Field Bean Scheme with standards of 0 and 2 infected seeds per 1000 for the basic and certified status, respectively, and less than 1% for the commercial status, has kept the incidence of the disease at minimal levels (Hewett, 1973). The Canadian Seed Growers Association adopted tolerance levels of 0.75% or less infected seeds for the certified status, 0.2% or less for the foundation status, no infected seeds per 1000 for the select status and 0.25% or less for imported seed lots (Bernier, 1975). The application of these standards checked the disease effectively in Canada. Liew and Guant (1980) and Liew (1981) found reduced disease incidence in the field with benomyl and RH 2161 (Rhom and Haas Ltd.) slurry seed treatments.

Recently Robertson and Hanounik (1987) have used the Ascochyta blight-resistant lines BPL471, 646, 74 and 2485 (Hanounik and Robertson, 1988) and combined resistance with high yield and alternative plant growth habit. Bond *et al.* (1987) have used the winter-hardy resistant line IB18-1/3 for production of composite cultivars in the UK.

Stem nematodes

Distribution and importance. The stem nematode *Ditylenchus dipsaci* is a devastating soil- and seed-borne nematode of faba bean in many parts of the temperate region (Hooper, 1983). Infested seed plays an important role in the survival and dissemination (Hooper, 1971) of the nematode and this is probably why *D. dipsaci* has a very wide geographical distribution (Hooper, 1983).

Although various biological races have been reported in stem nematode (Seinhorst, 1957), faba bean is attacked only by the "giant race" and the "oat race" (Hooper, 1983). The "giant race" is generally more common in the Mediterranean region (Hanounik and Sikora, 1980; Lamberti, 1981) compared to the "oat race" in Europe (Hooper, 1983). The "giant race" is responsible for greater damage and higher percentage of infested seed, compared to the "oat race" (Hooper and Brown, 1975). In Syria, faba bean yield was reduced by 68%, with 20% of the seed infested, in experimental plots containing 650 larvae of the "giant race" per 100 m^n soil (Hanounik, 1983).

Epidemiology. Reports on the epidemiology of *D. dipsaci* on faba bean are scanty. Stem nematodes have an outstanding ability to survive anhydrobiosis in seed and plant debris. The "giant race" of *D. dipsaci* remained pathogenic after three years of storage of dry infested seed in the laboratory in Syria (Hanounik, 1983). The nematode can also survive passage through animal digestive systems (Marinari *et al.*, 1971). The life cycle of *D. dipsaci* can be completed in 19-23 days at 21°C (Yuksel, 1960) and three to four generations may occur in one season. Each female lays as many as 500 eggs and one infested seed may contain 100-4000 nematodes (Ighil and Caubel, 1986).

Control. There is no single method to control *D. dipsaci*. Although the use of clean seed and nematode-free sites reduce disease incidence, effective management can be achieved only if host resistance and chemicals are integrated with other control measures.

Evaluation of ICARDA's pure line collection of faba bean in 1981 and 1982 in artificially infested soil, in the field at the Lattakia sub-site in northern Syria, led to the identification of 12 lines resistant to stem nematode (Hanounik *et al.*, 1986). Resistance in these lines was reconfirmed in field screening in Tunisia in 1984 and in France in 1990.

Winfield (1971) controlled *D. dipsaci*-infected narcissus bulbs by a hot water dip at 44-45°C for 3-4 hours. Hot water treatment should be evaluated to control seed-borne *D. dipsaci* in faba bean.

Methyl bromide is highly effective against free larvae of *D. dipsaci*, but dosages needed for complete eradication of the eelworm from infected seed cause a substantial decrease in germination (Powell, 1974). The effect of aldicarb (Temik-10 G) on faba bean yield and seed-infestation was studied at population densities ranging between 0 and 6500 larvae of *D. dipsaci* per 1000 cm^n soil in the field near Lattakia in Syria. Results showed that yield was reduced significantly as the population density of *D. dipsaci* increased beyond 40 larvae/1000 cm^n soil. Soil treatment with aldicarb increased yield at all population densities compared to untreated plots. The use of aldicarb gave nematode-free seed at all population densities except at the highest density of 6500 larvae/1000 cm^n soil. Plants grown in untreated plots, however, yielded infested seeds even at the lowest population density of 40 larvae/1000 cm^{n} soil. The percentage of infested seed increased from 5% with the lowest population density to 20% with highest population density of *D. dipsaci* in the soil (Hanounik, 1983).

Lentil

Vascular wilt

Distribution and importance. Vascular wilt caused by *Fusarium oxysporum* f.sp. *lentis* is an important and widely distributed disease of lentil. The disease has been reported from various areas around the world, including South America (Ravenna and Negri, 1979), USA (Kaiser, 1981), Canada (Bhalla *et al.*, 1984), Europe (Moreau, 1978), North Africa (Djerbi *et al.*, 1978), the Middle East (Bayaa *et al.*, 1986) and India (Khare, 1981). Vascular wilt is considered as an important disease in India causing more than 50% loss in some fields (Khare, 1981).

Epidemiology. Information on the epidemiology of vascular wilt of lentil is fragmentary. Optimum temperature for fungal growth is between 22°C and 25°C (Dhingra *et al.*, 1974). Fusarium wilt can be destructive between 17°C and 31°C, particularly in sandy loam soil with about 25% moisture (Khare, 1981). Early sowings are more favourable for disease development than late sowing (Kannaiyan and Nene, 1975). *F. oxysporum* f.sp. *lentis* is mainly a soil-borne pathogen and its host range is restricted to lentil. Seed transmission of the pathogen is still controversial. Although the pathogen has been isolated from lentil seed (Khare, 1981), there is no evidence on the relationship between seed infection and disease incidence in the field. Erskine *et al.* (1990) failed to detect the fungus in seed harvested from diseased plants.

Control. Sources of resistance have been identified in India (Kannaiyan and Nene, 1976; Khare and Sharma, 1969) and in Syria (Bayaa and Erskine, 1990).

Various cultural and sanitary practices including date of planting, long rotation, use of clean seed and early maturing varieties are helpful in reducing disease incidence. Chemical seed treatment reduces disease incidence (Khare *et al.*, 1974).

Ascochyta blight

Distribution and importance. Ascochyta blight caused by *Ascochyta lentis* is a serious and widely distributed disease of lentil. The disease has been reported from several areas of the world including South America, USA, Europe (Kaiser and Hannan, 1986) and the Middle East (Bellar and Kebabeh, 1983). The disease also occurs in India, Pakistan and the USSR. Low seed quality due to the disease is probably more important than decreased yield (Mathur *et al.*, 1988).

Epidemiology. The epidemiology of blight in lentil is not well understood. The pathogen is both stubble- and seed-borne. Kaiser and Hannan (1986) examined 284 seed samples from 30 countries and detected the pathogen in 16% of samples originating from 16 countries. Frequency of transmission from infected seed to seedling is low, especially when soil temperature is moderate to high. The progress of foliar blight is rapid and can lead to epidemic proportions under cool and wet weather conditions (Slinkard *et al.*, 1983).

Control. Resistant sources to *A. lentis* have been identified (Erskine, 1985; Singh *et al.*, 1982).

Seed treatment with thiabendazole, etaconazole, captan, benomyl and calixin-M, provided good control *in vitro*, and reduced disease incidence in the field (Bashir *et al.*, 1986).

One foliar application of chlorothalonil, captafol, folpet or metiram at early bloom to early pod set provided good protection (Beauchamp *et al.*, 1986). Using clean seed and ploughing down infected debris are recommended. Sun drying of lentil seed with polyethylene sheets may provide up to 96% control, but this method reduced seed germination (Beniwal *et al.,* 1989).

Rust

Distribution and importance. Rust of lentil caused by *Uromyces fabae* was first reported from Cyprus (Nattrass, 1932). It is the most devastating foliar disease of lentil. The disease has been reported from Iran, India, Bulgaria, Palestine, Morocco, Portugal, Turkey and Sicily (Khare, 1981). Losses of up to 100% have been reported from India (Khare and Agrawal, 1978). The disease has recently become a limiting factor in lentil production in North Africa.

Epidemiology. *U. fabae* is autoecious and completes its life cycle on lentil. Infected plant debris mixed with seed can play an important role in the recurrence of the disease (Prasad and Verma, 1948). Evidence for true seed transmission is lacking (Richardson, 1979). Disease development is rapid at 20-22°C, particularly when humid, cloudy and rainy weather conditions prevail.

Control. The use of rust resistant varieties is the most efficient and cheapest method to control the disease. Some sources of resistance have been identified (Agrawal *et al.*, 1976b). Two lines from ICARDA which are resistant to the disease have been released for their high performance (ILL358 in Ethiopia, and ILL4050 "Centinela + INIA" in Chile (Bascur and Sepulveda, 1989). Recently five lines from ICARDA (i.e. ILL 215, 255, 234, 275 and 277) were found highly resistant to rust in Morocco.

The use of clean, certified seed, and the destruction of infected plant debris reduce chances of primary infection.

Foliar sprays with Dithane M-45 or ferbam (Agrawal *et al.*, 1976a) were effective.

References

Agrawal, S.C., Khare, M.N. and Agrawal, P.S. 1976a. Control of lentil rust by use of fungicides. *Indian Phytopathol.* **29**: 90-91.

Agrawal, S.C., Khare, M.N. and Agrawal, P.S. 1976b. Field screening of lentil lines for resistance to rust. *Indian Phytopathol.* **29:** 208.

Bascur, B.C. and Sepulveda, R.P. 1989. Centinela-INIA. A new lentil variety tolerant to rust (*Uromyces fabae*). *Lens Newsletter* **16**: 24-26.

Bashir, M., Alam, S.S. and Malik B.A. 1986. *In vitro* evaluation of fungicides against *Ascochyta lentis*. *Lens Newsletter* **13**: 26-28.

Bayaa, B., Erskine, W. and Khoury, L. 1986. Survey of wilt damage on lentils in Northern Syria. *Arab Journal of Plant Protection* **4**: 118-119.

Bayaa, B. and Erskine, W. 1990. A screening technique for resistance to vascular wilt in lentil. *Arab Journal of Plant Pathology* **8**: 30-33.

Beauchamp, C.J., Morrall, R.A.A. and Slinkard, A.E. 1986. The potential for control of *Ascochyta* blight of lentil with foliar applied fungicides. *Canadian Journal of Plant Pathology* **8**: 254-259.

Bellar, M and Kebabeh, S. 1983. A list of diseases, injuries, and parasitic weeds of lentils in Syria, Survey 1979-1989. *Lens Newsletter* **10**: 30-31.

Beniwal, S.P.S., Ahmed, S. and Tadesse, N. 1989. Effect of sun drying of lentil seeds on the control of seed-borne *Ascochyta lentis*. *Lens Newsletter* **16**: 27-28.

Bernier, C.C. 1975. *Diseases of pulse crops and their control oil seed and pulse crops in W. Canada.* pp. 439-454. (ed. J.T. Harapiek) Western Cooperative Fertilizers, Ltd., Calgary, Alberta.

Bhalla, M.K., Nossolillo, C. and Schneider, E.F. 1984. Pathogenicity of soil fungi associated with root rot of lentils. *Canadian Journal of Plant Pathology* **6**: 21-28.

Bond, D.A., Brown, S.J., Jellis, G.L., Pope, M., Hall, J.A. and Clarke, M.H.E. 1987. *Ascochyta*. pp. 45-46. In: *Annual Report of the Plant Breeding Institute for 1986*. Cambridge, UK.

Dhingra, O.D., Agrawal, S.C., Khare, M.N. and Kushawa, L.S. 1974. Temperature requirements of eight strains of *Fusarium oxysporum* f.sp. *lentis* causing wilt of lentils. *Indian Phytopathology* **27**: 408-410.

Djerbi, M., Mlaiki, A. and Bouslama, M. 1978. Food legume diseases in North Africa. pp. 103-105. In: *Food Legume Improvement and Development* (eds. G.C. Hawtin and G.J. Chancellor) International Center for Agricultural Research in the Dry Areas and International Development Research Center.

Dodd, I.J. 1971. Some aspects of the biology of *Ascochyta fabae* Speg. (leaf and pod spot of the field bean, *Vicia fabae* l.). Ph.D. thesis, University of Hull, England.

Erskine, W. 1985. *Perspective in lentil breeding in faba beans, Kabuli chickpeas, and lentils in the 1980s*. pp. 19-100. ICARDA, Syria.

Erskine, W., Bayaa, B. and Dolli, M. 1990. Effect of temperature and some media and biotic factors on the growth of *Fusarium oxysporum* f.sp. *lentis* and its mode of seed transmission. *Arab Journal of Plant Protection* **8**: 34-37.

Guant, R.E. 1983. Shoot diseases caused by fungal pathogens. pp. 463-492. In: *The Faba Bean (Vicia faba L.). A basis for improvement.* (ed. P.D. Hebblethwaite). Butterworths, London.

Guant R.E. and Liew, R.S.S. 1981. Control strategies for *Ascochyta fabae* in New Zealand field and broad bean crops. *Seed Science and Technology* **9**: 707-715.

Guant, R.E., Teng, P.S. and Newton, S.D. 1978. The significance of *Ascochyta* leaf and pod spot disease in field bean (*Vicia faba* L.) crops in Canterbury, 1977-78. *Proc. Agron. Soc. N.Z.* **8**: 55-57.

Hanounik, S.B. 1979. Diseases of major food legumes in Syria. In: *Food Legume Improvement and Development*. (eds. G.C. Hawtin and G.J. Chancellor). IDRC Pub. 12e, Ottawa.

Hanounik, S.B. 1981. Influence of "Ronilan" on the severity of chocolate spot and yield of faba bean. *FABIS* **3**: 50-51.

Hanounik, S.B. 1983. Effects of aldicarb on *Ditylenchus dipsaci* in *Vicia faba*. pp. 1149. In: *Proceedings of the 10th International Congress of Plant Protection*. November 22-25, 1983. Brighton, UK.

Hanounik, S.B. 1988. Biological control of *Botrytis fabae. Report of the Food Legume Improvement Program for 1988.* pp. 46-48. ICARDA, Aleppo, Syria.
Hanounik, S.B. and Hawtin, G.C. 1982. Screening for resistance to chocolate spot caused by *Botrytis fabae*. In: *Faba Bean Improvement.* (eds. G.C. Hawtin and C. Webb). Martinus Nijhoff publishers, The Netherlands. 298 pp.
Hanounik, S.B. and Maliha, N. 1986. Horizontal and vertical resistance to chocolate spot caused by *Botrytis fabae. Plant Disease* **70**: 770-773.
Hanounik, S.B. and Robertson, L.D. 1987. New sources of resistance in *Vicia faba* to chocolate spot caused by *Botrytis fabae. Plant Disease* **72**: 696-698.
Hanounik, S.B. and Robertson, L.D. 1988. Resistance in *Vicia faba* germplasm to blight caused by *Ascochyta fabae. Plant Disease* **73**: 202-205.
Hanounik, S.B. and Sikora, R.A. 1980. Report of stem nematode *Ditylenchus dipsaci* on faba bean in Syria. *FABIS* **2**: 49.
Hanounik, S.B., Halila, H. and Harabi, M. 1986. Resistance in *Vicia faba* to stem nematodes *Ditylenchus dipsaci. FABIS* **16**: 37-39.
Harrison, J.G. 1978. Role of seed borne infection in epidemiology of *Botrytis fabae* on field beans. *Transactions of the British Mycological Society* **70**: 35-40.
Harrison, J.G. 1979. Overwintering of *Botrytis fabae. Transactions of the British Mycological Society* **72**: 389-394.
Hewett, P.D. 1973. The field behaviour of seed-borne *Ascochyta fabae* and disease control in field beans. *Annals of Applied Biollogy* **74**: 287-295.
Hooper, D.J. 1971. Stem eelworm *Ditylenchus dipsaci*, a seed and soil borne pathogen of field beans *Vicia faba. Plant Pathology* **20**: 25-27.
Hooper, D.J. 1983. Nematode pests of *Vicia faba* L. pp. 347-370. In: *The Faba Bean* (ed. P.D. Hebblethwaite) Butterworths, London U.K.
Hooper, D.J. and Brown, G.J. 1975. Stem nematode *Ditylenchus dipsaci*, on field beans. pp. 183-184. Rothamsted Exp. Sta. Dept. 1974. Part I.
Ighil, M.A. and Caubel, G. 1986. Contamination des graines de *Vicia faba* par le nématode des tiges, *Ditylenchus dipsaci.* Conséquences épidémiologiques. *Seed Science and Technology* **14**: 431-438.
Ingram, J. and Hebblethwaite, P.D. 1976. Optimum economic seed rates in spring and autumn sown field beans. *Agric. Prog.* **51**: 27-32.
Kaiser, W.J. 1981. Diseases of chickpea, lentil, pigeon pea, and tepary bean in continental United States and Puerto Rico. *Economic Botany* **35**: 300-320.
Kaiser, W.J. and Hannan, R.M. 1986. Incidence of seed-borne *Ascochyta lentis* in lentil germplasm. *Phytopathology* **76**: 355-360.
Kannaiyan, J. and Nene, Y.L. 1975. Note on the effect of sowing date on the reaction of 12 lentil varieties to wilt disease. *Madras Agricultural Journal* **2**: 77-79.
Kannaiyan, J. and Nene, Y.L. 1976. Reaction of lentil germplasm and cultivars against three root pathogens. *Indian Journal of Agricultural Sciences* **46**: 165-167.
Khare, M.N. 1981. Diseases of lentils. pp. 163-172. In: *Lentils* (eds. Webb, C. and Hawtin, G.C.) CAB and ICARDA.
Khare, M.N. and Agrawal, S.C. 1978. *Lentil Rust Survey in Madhya Pradesh.* All India Workshop held at Barada, 3pp.
Khare, M.N., Agrawal, S.C., Kushwaha, L.S. and Tomar, K.S. 1974. Evaluation of fungicides for the control of wilt of lentil caused by *Sclerotium rolfsii. Indian Phytopathology* **27**: 364-366.
Khare, M.N. and Sharma, H.C. 1969. Screening of lentil varieties against *Fusarium* wilt. *Mysore Journal of Agricultural Sciences* **4**: 354-256.
Lamberti, F. 1981. Plant nematode problems in the Mediterranean region. Helminthological Abstracts, Series B, *Plant Nematology* **50**: 145-166.
Liew, R.S.S. 1981. The epidemiology, physiology and control of *Ascochyta fabae* Speg. on *Vicia faba* L. Ph.D. thesis, Lincoln College, New Zealand.
Liew, R.S.S. and Guant, R.E. 1980. Chemical control of *Ascochyta fabae* in *Vicia faba. New Zealand Jl. Exp. Agric.* **8**: 67-70.

Marinari Palmisano, A., Tacconi, R. and Canestri Trotti, G. 1971. Sopravvivenza di *Ditylenchus dipsaci* (Kuhn) Filipjev (Nematoda: Tylenchidae) al processo digestivo nei suini, equini e bovini. *Redia* **52**: 725-737.

Mathur, S.B., Haware, M.P. and Hampton, R.O. 1988. Identification, significance and transmission of seed-borne pathogens. pp. 351-365. In: *World Crops: Cool Season Food Legumes* (ed. R.J. Summerfield).

Moreau, B. 1978. Maladies et insectes de la lentille, en les légumes secs, lentille vert, haricot flageolet, pois de casserie. pp. 95-100. Institut National de Vulgarisatio pour les fruits, légumes et champignons, Paris, France.

Nattrass, R.M. 1932. Annual Report, Department of Agriculture, Cyprus, pp. 56-64.

Powell, D.F. 1974. Fumigation of field beans against *Ditylenchus dipsaci*. *Plant Pathology* **23**: 110-113.

Prasad, R. and Verma, U.N. 1948. Studies on lentil rust, *Uromyces fabae*. *Indian Phytopathology* **1**: 142-146.

Ravenna, N.A. and Negri, R.C. 1979. Marchitamiento de la lenteja provocado por *Fusarium* sp. en la zona de Rosario Universidad Nacional de Tucuman, Facultad de Agronomia y Zootecnía, 3 Jornadas fisotanitarias argentinas. pp. 759-765.

Richardson, M.J. 1979. *An Annotated List of Seed-borne Diseases*. 3rd edition. ISTA Zurich, Switzerland. 124 pp.

Robertson, L.D. and Hanounik, S.B. 1987. Development of improved cultivars and genetic stocks for assured moisture environments; disease resistance. *Annual Report of the Food Legume Improvement Program for 1987*. pp. 24-25. ICARDA, Aleppo, Syria.

Seinhorst, J.W. 1957. Some aspects of the biology and ecology of stem eelworms. *Nematologica* **2**: 355-361.

Singh, G., Singh, K., Gill, A.S. and Brar, J.S. 1982. Screening of lentil varieties and lines for blight resistance. *Indian Phytopathology* **35**: 678-679.

Slinkard, A.E., Morrall, R.A.A. and Gossen, B. 1983. *Ascochyta lentis* on lentils in Canada, 1982. *Lens Newsletter* **10**: 31.

Sode, J. and Jorgensen, J. 1974. Correlation between the incidence of disease on seed of horse beans, on the plants in the field and in the harvested seeds. *Beretning, Statsfrøkontrollen, Danmark* **103**: 99-106.

Sundheim, L. 1973. *Botrytis fabae*, *B. cinerea* and *Ascochyta fabae* on broad bean *Vicia faba* in Norway. *Acta. Agr. Scan.* **23**(1): 43-51.

Winfield, A.L. 1971. Control of bulb scalemite and stem nematodes of narcissus and reclaiming forced bulbs. *Plant Pathology* **20**: 10-13.

Yuksel, H.S. 1960. Observation on the life cycle of *Ditylenchus dipsaci* on onion seedlings. *Nematologica* **5**: 289-296.

Chickpea and Groundnut Seed-borne Diseases of Economic Importance: Transmission, Detection and Control

M.P. HAWARE, D.V.R. REDDY and D.H. SMITH

International Crops Research Institute for the Semi-Arid Tropics (ICRISAT), Patancheru, Andhra Pradesh 502 324, India

Introduction

With the establishment of International Agricultural Research Centers (IARC) the international exchange of seed has greatly increased. For example, during 1989 ICRISAT exported 13164 samples of chickpea seed and 6095 samples of groundnut seed to 41 and 37 countries, respectively. There is always a risk that pathogens can be transmitted by seed.

Seed-borne fungi, bacteria and viruses result in yield losses, reduction in seed germination, increased risk of deterioration in storage, and harmful effects on humans and animals because of toxic metabolites of certain mold fungi.

We need to approach the problem of seed-borne diseases rationally and develop a research programme that is scientifically sound. Not all microorganisms carried in or on the seed are seed-borne pathogens. Before developing a disease management programme it is important to know which of the seed-borne diseases are capable of causing reduction in yield and quality and what is the importance of seed-borne inoculum in spread of the disease.

In this paper we describe the more important seed-borne diseases of chickpea and groundnut, and discuss their transmission, detection, and control.

Seed-borne diseases of chickpea caused by fungi

Chickpea (*Cicer arietinum*) is an important grain legume crop in the Indian subcontinent, Middle East, Northern and Eastern Africa and Central and South America. It is an important source of protein. According to the FAO Production

---

Submitted as Conference Paper No. 694 of the International Crops Research Institute for the Semi-Arid Tropics (ICRISAT), Patancheru, Andhra Pradesh 502 324, India.

Year Book (1988), chickpea was planted on an area of 10.61 million hectares in 1986 and production was nearly 7 million tonnes. Though the overall yield potential of present day cultivars is over 3000 kg ha^{-1}, the average productivity is only about 700 kg ha^{-1}.

About 45 fungal diseases have been reported on chickpea from different parts of the world (Nene *et al.*, 1989). Of economic importance are Fusarium wilt (*Fusarium oxysporum* f.sp. *ciceri*), dry root rot (*Rhizoctonia bataticola*), Ascochyta blight (*Ascochyta rabiei*) and Botrytis gray mold (*Botrytis cinerea*). Of the numerous diseases recorded, very few are seed-borne. Haware *et al.* (1978) described the seed-borne nature of *F. oxysporum* f.sp. *ciceri*, and seed-borne diseases caused by this species and by *A. rabiei*, *B. cinerea*, *Colletotrichum dematium* and *Alternaria alternata*. Methods for their detection in seed were described in a technical bulletin (Haware *et al.*, 1986). This bulletin is primarily intended for seed production, seed certification, and plant quarantine personnel.

Fusarium wilt

The disease has been reported from several countries. Early infection sometimes kills plants, resulting in total yield loss. In India, it is estimated to cause a 10% annual yield loss. The fungus is a vascular pathogen and is soil- and seed-borne. Seed-borne infection is usually present in seeds harvested from plants which wilt after pod formation. Seeds from wilted plants are generally small, wrinkled, and discoloured. Thus diseased seed can be detected visually, but a seemingly normal seed may also harbour the pathogen. Therefore, it is important to test the seed for the presence of the fungus. Haware *et al.* (1978) showed that the fungus was present in the hilum region of the seed in the form of chlamydospore-like structures. These structures were thick-walled, spherical, closely packed, and connected by hyphal cells. Chickpea cultivars differ in the extent of yield loss and seed infection (Haware and Nene, 1980). The most common methods of pathogen dispersal are apparently by seed and soil.

For detection, 400 seeds are surface-sterilized by immersing them for 2 min in 2.5% aqueous solution of sodium hypochlorite. Seeds are then plated onto modified Czapek-Dox agar (10 per plate) and incubated at 20°C for 8 days in a diurnal cycle of 12 h of near-UV light followed by 12 h darkness. The white cottony mycelium of *F. oxysporum* f.sp. *ciceri* can be observed emerging from the seed (Haware *et al.*, 1986). A seedling symptom test should be employed if an agar medium is not available. Surface-sterilized seeds are sown in soil or fine riverbed sand in pots. These pots are kept in a growth chamber or in a glasshouse at 25°C in a diurnal cycle of 12 h light and 12 h darkness. Seedlings should be monitored for wilt symptoms until at least 40 days after sowing (DAS). The seedlings from infected seeds generally wilt between 15 and 25 DAS. The fungus can be isolated from roots. The wilt count closely agrees with the number of colonies detected on selective medium (Haware *et al.*, 1978).

Ascochyta blight

It is an important disease of chickpea in Pakistan, West Asia, and Northern Africa. In Pakistan, about 70% of the crop was lost to the disease in 1979 and in 1980 (Nene, 1982). It also appeared in epiphytotic form in parts of Punjab and Haryana States of India during 1980 and 1981. For the first time, in 1983, Kaiser and Muehlbauer (1984) reported a trace to a high incidence of blight in germplasm evaluation trials at Pullman, USA. Seventy-seven of 125 accessions tested were affected. Cool, wet weather during June and July favoured infection and spread of the pathogen. According to these authors, the pathogen was introduced into the USA on seed imported from Syria and/or India.

The most common and effective method of dissemination of *A. rabiei* appears to be in seed. Infected seeds are small, wrinkled, and have dark brown lesions of various shapes and sizes on their testae. Pycnidia are found in deep lesions on these seed. If pods are infected at maturity, a seemingly normal seed may show only slight discoloration on the surface but may harbour the pathogen.

For seed health testing, potato dextrose agar containing 1 g Dicrysticin-S per litre of medium is suitable. Seeds must be surface-sterilized by soaking them in a 2.5% aqueous solution of sodium hypochlorite for 2 min. *A. rabiei* is slow-growing and if surface contaminants on the seed are not killed, the pathogen may not be detected. Petri plates, each containing 10 seeds, are incubated in diurnal cycles of 12 h near-UV light and 12 h darkness at 22°C for 10 days. The colonies of the fungus grow slowly on seed and are creamy white with black centres.

In the seedling symptom test, seedling emergence is not necessarily affected by seed infection. Indeed, the test does not give a reliable estimate of seed infection because, in many seedlings, the emerging shoots escape fungal contact and thereby no infection of *A. rabiei* is detected.

Control

Host resistance. Sources of high level resistance to the wilt disease are available. Some of these genotypes have resistance to other diseases including dry root rot, black root rot, Botrytis gray mold, Ascochyta blight or Sclerotinia blight. Cultivars including Avrodhi, JG 315, BG 2344 and ICCC 32, ICCV 2 and ICCV 10 are resistant to wilt disease.

For Ascochyta blight, resistance sources both in kabuli and desi germplasm have been identified. Blight resistant cultivars (ILC 3279 in Syria; C 543 and G 688 in India; CM 72 in Pakistan) have been released in different chickpea growing countries. Because of pathogen variability in *A. rabiei*, resistant cultivars are susceptible to some populations of the pathogen.

Cultural practices. Since the wilt pathogen survives in soil for more than 6 years, crop rotation is not effective for disease control. Intercropping, plant population levels, and fertilizer application had no effect on wilt incidence (Zote *et al.*, 1986).

In seeds from plants which wilt after pod formation, seed-borne infection is commonly observed. Hence roguing of wilted plants (at least for seed purposes) at harvest reduces the number of infected seeds. Use of inoculum-free seed reduces the probability of spreading the pathogen to new areas.

A. rabiei can survive in diseased crop debris on or near the soil surface for two years (Luthra *et al.*, 1935). Crop rotation can eliminate this source of primary inoculum. In order to neutralize this inoculum source the dead plant debris should be removed or ploughed deep. This would probably eliminate the production of the perfect stage of the pathogen in the overwintered crop debris. Viable ascospores can be dispersed for long distances on air currents and thus may introduce the pathogen to new areas (Trapero-casas and Kaiser, 1987). Sexual reproduction could also contribute to increased variability of the pathogen.

Seed treatment. An 0.1% Ceresan[R] solution seed treatment followed by a seed treatment with 0.2% thiram or 0.2% PCNB has been reported to suppress wilt development. Bavistin[R] seed treatment at 2.5 g kg^{-1} seed rate was also effective in pot experiments (Shukla *et al.* 1981). Seed dressing with Benlate T[R] (30% benomyl + 30% thiram) at 1.5 g kg^{-1} of seed eradicated the seed-borne inoculum (Haware *et al.*, 1978) of chickpea fusarium wilt. A Calixin M[R] (11% tridemorph + 36% maneb) seed treatment has eradicated the seed-borne inoculum of *A. rabiei* (Reddy *et al.*, 1982). The slurry seed treatment at 2.5 g kg^{-1} with Calixin M[R], Calixin M[R] + thiram (1:1), Calixin M[R] + Bavistin[R], and Bavistin[R] + thiram (1:3) when tested by the blotter-method apparently eliminated the seed-borne inoculum. But when plated on Oat-meal agar, the fungus was isolated from 4-6 percent of treated seed (*A. rabiei*) as compared with 30% recovery from untreated seed. However, Calixin M[R], Calixin M[R] + thiram, and Calixin M[R] + Bavistin[R] treatments significantly reduced seedling vigor. Bavistin[R] + thiram seed treatment was best for eradication of seed-borne inoculum, significantly increasing the seed germination and seedling vigor (Tripathi *et al.*, 1986). Thiabendazole seed treatment (3 g ka^{-1} seed) is more effective and safer than using Calixin M[R] (Reddy and Kababeh, 1984). Treatment of seed with an effective fungicide would control the disease most economically allowing free international distribution of seed and reducing the danger of introducing the pathogen into new areas.

Seed-borne diseases of groundnut caused by fungi, nematodes, and bacteria

The groundnut or peanut (*Arachis hypogaea*) was originated in South America, but it is now widely cultivated in tropical and subtropical areas of six continents. The groundnut has high oil and protein contents. Haulms are fed to domestic animals in many countries.

A world list of groundnut diseases was recently published by Subrahmanyam *et al.*, (1990). There are over 60 fungal diseases, 17 viral diseases, one bacterial disease, one disease caused by a Rickettsia-like organism, one disease caused by a mycoplasma-like organism, 10 nematode diseases, and two phanerogamic parasites.

Numerous fungi have been isolated from groundnut seed. The fungi frequently associated with seed rotting and seedling diseases are: *Aspergillus flavus*, *Aspergillus niger*, *Fusarium* spp., *Lasiodiplodia threobromae*, *Macrophomina phaseolina*, *Penicillium* spp., *Pythium* spp., *Rhizopus* spp., *Sclerotinia minor* and *Sclerotium rolfsii*. Most of these pathogens occur widely in many groundnut-growing areas of the world.

Several seed protectant fungicides are used either as a single fungicide or as mixtures of two or more fungicides, depending on the spectrum of fungal pathogens in different groundnut production areas of the world. Formulations of captan, carboxin, DCNA, ethazol, maneb, Metalaxyl, PCNB and thiram are recommended for application to groundnut seed.

Most groundnut seed-rotting pathogens can be detected by plating surface sterilized seeds on potato dextrose agar, but selective methods are available for isolation of species of *Aspergillus*, *Fusarium*, *Macrophomina*, and *Fusarium*.

Sclerotinia minor, the causal agent of Sclerotinia blight of groundnuts, has a wide host range (Akem and Melouk, 1990; Porter *et al.*, 1989). However, since the current distribution of Sclerotinia blight of groundnut is limited, the distribution of seed produced in fields infected with *S. minor* should be avoided. Sclerotia of *S. minor* commonly form within pods and occasionally within seeds. Therefore, seed protectant fungicides may not be effective. Sclerotia of *S. minor* can survive for several years in groundnut fields. This is another reason for avoiding distribution of seed produced in infested fields.

Ditylenchus destructor is an important nematode of groundnut in South Africa (De Waele *et al.*, 1989). The nematode attacks pegs, pods, and seeds. Infested seeds are shrunken. Micropyles are dark brown to black. Testae are flaccid and easily removed, and vascular strands in the testa are dark coloured. The inner layer of the testa has a yellow discoloration. Infected embryos are usually olive green to brown. Detection of this nematode can be done by placing groundnut seed in water for 24 hours. The nematodes can then be observed with a stereoscopic microscope.

We are not aware of measures that can be used to control this nematode either within or on groundnut seeds.

Mahmud and Middleton (1990) reported seed transmission of *Psuedomonas solanacearum* the causal agent of bacterial wilt. Some harvested groundnut pods from infected plants were discoloured and some pod rot was observed. No symptoms were observed on pods collected from healthy plants. Discoloration of the seed coat and cotyledon was observed on some seed obtained from infected plants. Discoloration of the embryo was rarely observed. When infected seeds were planted, the incidence of wilting was 5 to 8% at 2 to 4 weeks after planting.

P. solanacearum can be detected using the SPA medium (Hayward, 1964) or the TZC medium (Kelman, 1954). Colonies of bacterium are fluidal and irregular in shape and white or pink centres that darken with age. Avirulent colonies are round butyrous and uniformly red even at the early stage of growth.

Seed-borne virus diseases of groundnut and chickpea

Several economically important groundnut diseases are caused by seed-borne viruses. However, only one seed-borne virus of minor importance has been reported on chickpea (Kaiser *et al.*, 1990). Of the plant viruses which infect groundnut under natural conditions (Reddy, 1991) at least five are seed-transmitted (Tables 1 and 2). Infection of gametes is necessary to facilitate seed transmission of viruses. In the case of peanut mottle (PMV) and peanut stripe (PStV) viruses, infected seed serve as the primary source of inoculum.

Despite availability of highly sensitive and reliable methods for detection of seed-borne viruses, these techniques are not widely used in developing countries. Thus in many countries these viruses may not be detected with current inspection methods. Seed-borne viruses can be detected either by a direct method (growing-on tests) or by indirect methods (infectivity assays, serological tests, and complementary DNA probes). For growing-on tests seedlings are raised in sterilized soil in a greenhouse. Viruses which produce macroscopic symptoms can be easily recognized, but those that produce no overt symptoms escape in this test. Since diagnostic hosts are currently available for all seed-transmitted viruses in groundnut (Table 2), infectivity assays can be used for virus detection. They are recommended for use in countries where facilities for performing serological tests and nucleic acid hybridization tests are not available. Infectivity assays are especially valuable for virus detection in plants exhibiting no symptoms in growing-on tests.

Of all serological tests currently available for the detection of groundnut viruses, enzyme-linked immunosorbent assay (ELISA) is the most preferred assay method. ELISA methods for detection of viruses in groundnut seed have been described (Sudarshana *et al.*, 1990).

All currently known seed-transmitted groundnut viruses contain single stranded RNA. It is possible to produce complementary DNA (cDNA) probes which can be used in a variety of nucleic acid hybridization tests. cDNA probes have been used for detecting PMV and PStV in groundnut seed (Bijaisoradat and Kuhn, 1988). Nucleic acid hybridization tests need highly perishable enzymes and elaborate laboratory facilities. Until non-radioactive cDNA probes (Roy *et al.*, 1988) are made available at affordable costs to research workers in developing countries cDNA probes will probably not be utilized for routine virus detection in seed.

Table 1. Features for identification of seed-transmitted viruses of groundnut.

Virus and virus group	Particle morphology	Serological reactions positive	negative
Peanut mottle potyvirus	Flexuous rods 740-750 nm length 13 nm width	Adzuki bean mosaic	Bean yellow mosaic, Groundnut eyespot, Peanut green mosaic, Peanut stripe
Peanut stripe potyvirus	Flexuous rods 752 nm length 13 nm width	Black eye cowpea mosaic Clover yellow vein Soybean mosaic	Bean yellow mosaic, peanut mottle
Peanut clump furovirus	Two rod shaped particles 245 nm length 160 nm length 24 nm width	Exist as several serologically distinct isolates. Cross reaction occurs among some isolates	Beet necrotic yellow vein, Potato mop top, Soil-borne wheat mosaic
Peanut stunt cucumovirus	Spherical 25-30 nm diameter	With several peanut stunt virus isolates from USA	-*
Cucumber mosaic cucumovirus	Spherical 28 nm diameter	Several isolates of cucumber mosaic and peanut stunt viruses	-*

* Data not given because of their limited value for identification.

The presence of serologically different isolates for viruses such as peanut clump (PCV) may pose problem for using serological techniques for virus detection in quarantine. Thus broad-spectrum monoclonal antibodies and cDNA probes have immense potential for virus detection in seed. Research on production of non-radioactive cDNA probes for PCV is in progress at ICRISAT. The two widely distributed seed-borne viruses, PMV and PStV, are potyviruses and it is possible that other potyviruses may transmit through groundnut seed. Thus it is also essential to produce antibodies which can detect a large number of potyviruses. Recently Shukla and Ward (1989) reported that polyclonal antibodies can be produced for the core region of viral polypeptide, which is highly conserved in different potyviruses. Research on production of polyclonal antisera, for the core region of potyvirus polypeptides, for detection of several potyviruses is in progress at ICRISAT.

Since control measures to eradicate the seed-borne virus inoculum are not available, it should be ensured that only virus-free seeds are used in germplasm exchange. Genotypes in which the virus is not seed-transmitted have been identified in case of PMV. Efforts are being made to develop agronomically acceptable, non-seed transmitting, high yielding breeding lines.

Table 2. Transmission frequency and host range of seed-transmitted viruses of groundnut.

Viruses	Percent seed Transmission	Diagnostic hosts Local lesions	Systemic
Peanut mottle virus	0.1 to 8.5	*Phaseolus vulgaris* (Topcrop)	*Glycine max* *Pisum sativum*
Peanut stripe virus	0.1 to 30	*Chenopodium amaranticolor* *C. quinoa*	*Glycine max* *Lupinus albus* *Vigna unguiculata*
Peanut clump virus	6.0 to 25	*Chenopodium quinoa* *Phaseolus vulgaris* *Vigna unguiculata* *N. glutinosa*	*Canavalia ensiformis* *Nicotiana clevelandii* *N. edwardsonii*
Peanut stunt virus	0.01 to 0.2	*Chenopodium amaranticolor* *Vigna unguiculata*	*Datura stramonium* *Glycine max* *Nicotiana tabacum*
Cucumber mosaic virus	1.0 to 2.0	*Chenopodium amaranticolor* *Datura stramonium* *Phaseolus mungo*	*Nicotiana tabacum* *Vigna unguiculata*

References

Akem, C.N. and Melouk, H.A. 1990. Transmission of *Sclerotinia minor* in peanut from infected seed. *Plant Disease* **74**: 216-219.

Bijaisoradet and Kuhn, C.W. 1988. Detection of two viruses in peanut seeds by complimentary DNA hybridization tests. *Plant Disease* **72**: 956-959.

De Waele, D., Jones, B.L., Bolton, C. and Van den Berg, E. 1989. *Ditylenchus destructor* in hulls and seed of peanut. *Journal of Nematology* **21**(1): 10-15.

F.A.O. 1988. *FAO Production Year Book 41*. Rome, Italy.

Haware, M.P. and Nene, Y.L. 1980. Influence of wilt at different growth stages on yield loss in chickpea. *Tropical Grain Legume Bulletin* **19**: 38-40.

Haware, M.P., Nene, Y.L. and Mathur, S.B. 1986. *Seed-borne Diseases of Chickpea*. Technical Bulletin No. 1, Copenhagen, Denmark: Danish Government Institute of Seed Pathology for Developing Countries, and Patancheru, A.P. 502 324, India: International Crops Research Institute for the Semi-Arid Tropics. 32 pp.

Haware, M.P., Nene, Y.L. and Rajeshwari, R. 1978. Eradication of *Fusarium oxysporum* f.sp. *ciceri* transmitted in chickpea seed. *Phytopathology* **68**: 1364-1367.

Hayward, A.C. 1964. Characteristics of *Pseudomonas solanacearum*. *J. Appl. Bacteriology* **27**(7): 265-277.

Kaiser, W.J., Ghanekar, A.M., Nene, Y.L., Rao B.S. and Anjaiah, V. 1990. Virus diseases of chickpea. "Chickpea in the Nineties". In: *Proceedings of the Second International Workshop on Chickpea Improvement*: 139-142.

Kaiser, W.J. and Muehlbauer, F.J. 1984. Occurrence of *Ascochyta rabiei* on imported chickpeas in eastern Washington. *Phytopathology* **74**: 1139 (Abstract).

Kelman, A. 1954. The relationship of pathogenicity in *Pseudomonas solanacearum* to colony appearance on a tetrazolium medium. *Phytopathology* **44**: 693-695.

Luthra, J.C., Sattar, A. and Bedi, K.S. 1935. Life history of gram blight (*Ascochyta rabiei*) on gram and its control in the *Punjab. Agr. Live-Stk. India* **5**: 489-498.

Mahmud, M. and Middleton, K.J. 1990. Transmission of *Pseudomonas solanacearum* through groundnut seeds. *FLCG Newsletter* **13**: 22-24.

Nene, Y.L. 1982. A review of ascochyta blight of chickpea. *Tropical Pest Management* **28**: 61-70.

Nene, Y.L., Sheila, V.K. and Sharma, S.B. 1989. World list of chickpea (*Cicer arietinum* L.) and pigeonpea (*Cajanus cajan* [L.] Millsp.) pathogens. *Legumes Pathology Progress Report* 7. ICRISAT, Patancheru, Andhra Pradesh 502 324, India. 23 pp.

Porter, D.M., Taber, R.A. and Smith, D.H. 1989. The incidence and survival of *Sclerotinia minor* in peanut seed. *Peanut Science* **16**: 113-115.

Reddy, D.V.R. 1991. Groundnut viruses and virus diseases: distribution, identification and control. *Review of Plant Pathology* **70:** 665-678.

Reddy, M.V. and Kababeh, S. 1984. Eradication of *Ascochyta rabiei* from chickpea seed with thiabendazole. *International Chickpea Newsletter* **10**: 17-18.

Reddy, M.V., Singh, K.B. and Nene, Y.L. 1982. Further studies of Calixin-M in the control of seed-borne infection of Ascochyta in chickpea. *International Chickpea Newsletter* **6:** 18-19.

Roy, B.P., Abowhaidu, M.G., Sit, T.L. and Alexander, A. 1988. Construction and use of cloned cDNA biotin and P^{32} labeled probes for the detection of papaya mosaic potexvirus RNA in plants. *Phytopathology* **78**: 1425-1429.

Shukla, D.D. and Ward, C.W. 1989. Structure of potyvirus coat proteins and its application in the taxonomy of the potyvirus group. *Advances in Virus Research* **36**: 273-314.

Shukla, P., Singh, R.R. and Mishra, A.N. 1981. Search for best seed dressing fungicides to control chickpea wilt (*Fusarium oxysporum*). *Pesticides* **15**: 15-16.

Subrahmanyam, P., Reddy, D.V.R., Sharma, S.B., Mehan, V.K. and McDonald, D. 1990. A world list of groundnut diseases. *Legumes Pathology Progress Report*. Patancheru, Andhra Pradesh 502324, India (ICRISAT). 14 pp.

Sudarshana, M.R., Reddy, D.V.R. and Reddy, A.S. 1990. Methods for the detection of seed-borne viruses in groundnut (*Arachis hypogaea* L.). In: *Proceedings of National Seminar on Advances in Seed Science and Technology*. (eds. H.S. Shetty and H.S. Prakash).

Trapero-Casas, A. and Kaiser, W.J. 1987. Factors influencing development of the teleomorph of *Ascochyta rabiei*. *International Chickpea Newsletter* **17**: 27-28.

Tripathi, H.S., Singh, R.S. and Chaube, H.S. 1986. Efficacy of some fungicides against *Ascochyta rabiei* (Pass) Labr. *International Chickpea Newsletter* **15**: 20-21.

Zote, K.K., Dandnaik, B.P., Raut, K.G., Deshmukh, R.V. and Katare, R.A. 1986. Effect of intercropping, plant population, fertilizer and land layouts on the incidence of chickpea wilt. *International Chickpea Newsletter* **15**: 16.

Seed Exchange

Seed Health Testing and Treatment of Germplasm at the International Center for Agricultural Research in the Dry Areas (ICARDA)*

MARLENE DIEKMANN

ICARDA, P.O. Box 5466, Aleppo, Syria

Introduction

One of the impacts of successful plant breeding programmes is the replacement of local varieties and/or land races by new, high yielding and disease and pest resistant varieties. While the introduction of new varieties on the farm helps to ensure food production in areas of increasing population, it simultaneously causes the disappearance of valuable land races and old varieties. This loss of genetic material provokes concern about the 'genetic erosion' of economically important crops.

Many institutions are now working worldwide on the collection and conservation of plant germplasm. Efforts are often coordinated and supported by the International Board for Plant Genetic Resources (IBPGR) and International Agricultural Research Centers (IARC's), which are largely financed through the Consultative Group on International Agricultural Research (CGIAR). The network of International Agricultural Centers seeks to increase the production of subsistence crops such as wheat (*Triticum aestivum*), maize (*Zea mays*), rice (*Oryza sativa*), potatoes (*Solanum tuberosum*), legumes, etc. Some Centers, such as ICRISAT [chickpea (*Cicer arietinum*), pigeon pea (*Cajanus cajan*)], CIMMYT (maize), CIP (potatoes), ICARDA [wheat, barley (*Hordeum vulgare*), lentils (*Lens culinaris*), large-seeded chickpeas and faba beans (*Vicia faba*), many forage legumes], are located near or at the 'center of origin' of the crops for which they have a mandate (CGIAR, 1980; Vavilov, 1951). They are actively involved in collecting diverse genetic material for use in crop improvement. At the Centers as well as at other institutions, large germplasm collections are stored under conditions favourable for maintaining seed viability over long periods.

There is, however, the risk of having in the stored germplasm a collection of various races of seed-borne pathogens which inadvertently could be spread to other countries. Different pathogens survive in or on seeds for various periods of time,

*Paper published in *seed Science and Technology* **16**: 405-417 (1988)

ranging from a few months to several years (Neergaard, 1977). Kaiser and Hannan (1986) isolated *Ascochyta lentis* from lentil germplasm that had been stored for more than 30 years. Generally, conditions which prolong seed viability also favour pathogen survival.

The standard procedures of seed health testing and seed treatment that are used for commercial seed lots can in principle be applied to germplasm. However, there are certain problems unique to the control of seed-borne pests and pathogens in germplasm collections and nursery materials. These problems will be highlighted and discussed in this paper.

ICARDA's germplasm exchange

Within the framework of the CGIAR, ICARDA has the world mandate for the improvement of barley, lentils and faba beans. It also has a regional mandate for the improvement of wheat, chickpeas and pasture and forage crops in a target area that extends from Morocco in the west to Pakistan in the east, and from Turkey in the north to Ethiopia in the south. The development of improved germplasm and new varieties for use by national, regional and international breeding programmes is the major objective of the ICARDA crop improvement programmes.

Seed exchange and unrestricted germplasm movement are vital for crop improvement in the ICARDA region, which is characterized largely by limited winter rainfall and hot dry summers. There are substantial parts of the region in high elevation areas where crops suffer from low temperature effects. Other stress factors that are serious constraints to production are heat, drought, and pests and pathogens. Scientists must exploit the genetic resources to develop germplasm targeted to many diverse agroecological zones. To avoid crop failures and disasters such as the epidemic of *Helminthosporium maydis*. in the USA in 1970/71, it is desirable that the genetic base of the crop be as broad as possible. This can be achieved by utilizing new germplasm of wild and related species and landraces.

Considerable resources are deployed to expand the genetic base of germplasm collections and to recombine various mechanisms of resistance to stress factors. In order to identify genetic material suitable for these various conditions, multilocation testing of germplasm is a prerequisite.

The collaborative work with plant scientists engaged in these breeding and testing programmes result in a massive flow of germplasm to and from ICARDA. More than 2000 sets of international nurseries are distributed every year. Each set consists of 24 entries (Cereals Yield Trials) to more than 500 entries (Key Location Disease Nurseries). In addition, numerous requests are met for specific selections from the breeding programmes or the germplasm collections.

Approximately 4000 different entries in more than 120,000 individual seed packets, each containing 10 to 100 g of seeds, are sent every year to cooperators in national programmes of the ICARDA region as well as to universities, breeding sations, etc., worldwide. A total of about 70 countries receive ICARDA's germplasm annually (Table 1).

Table 1. Countries which received germplasm from ICARDA (The International Center for Agricultural Research in the Dry Areas) during 1982-1985.

Afghanistan	Kuwait
Algeria	Lebanon
Argentina	Liberia
Australia	Libya
Austria	Malawi
Bahrain	Mali
Bangladesh	Mauritius
Belize	Mexico
Brazil	Morocco
Bulgaria	Nepal
Burma	Netherlands
Canada	Nigeria
Chile	Oman
China, People's Republic	Pakistan
Colombia	Peru
Costa Rica	Philippines
Cyprus	Poland
Czechoslovakia	Portugal
Ecuador	Qatar
Egypt	Saudi Arabia
Ethiopia	Somalia
Finland	Spain
France	Sri Lanka
Germany, Democratic Republic	Sudan
Germany, Federal Republic	Switzerland
Greece	Tanzania
Guatemala	Thailand
Honduras	Tunisia
Hungary	Turkey
India	United Arab Emirates
Indonesia	United Kingdom
Iran	United States of America
Iraq	USSR
Italy	Venezuela
Japan	West Indies
Jordan	Yemen, Arab Republic
Kenya	Yemen, People's Dem. Rep.
Korea	Zambia

ICARDA also receives seeds from different sources, such as research institutions and germplasm centres. The quantities can vary from a few seeds of wild species to several kilograms of commercial varieties. In 1985, more than 6000 different entries were obtained; most of these were dispatched from North African and West Asian countries (Table 2).

Table 2. Countries which sent germplasm to ICARDA (The International Center for Agricultural Research in the Dry Areas) during 1982-1985.

Algeria	Iraq
Argentina	Italy
Australia	Jordan
Bulgaria	Kenya
Canada	Lebanon
China, People's Rep.	Libya
Colombia	Mexico
Cyprus	Morocco
Czechoslovakia	Poland
Denmark	Portugal
Ecuador	Romania
Egypt	Spain
Ethiopia	Sudan
France	Switzerland
Germany, Democratic Rep.	Tunisia
Germany, Federal Rep.	Turkey
Greece	United Kingdom
Hungary	United States of America
India	USSR
Iran	Yugoslavia

Germplasm exchange and seed-borne pests and pathogens

A considerable number of pathogens are disseminated by seeds. Richardson (1979, 1981, 1983) listed some 500 different plant species that are attacked by more than 1300 different pathogens; many of these pathogens can attack more than one plant species. There are probably only very few plant species that are not hosts for one or more pathogens. About 80% of the seed-transmitted diseases are caused by fungi. The remaining 20% of pathogens are viruses and possibly viroids as well as bacteria and nematodes.

Often germplasm is stored for many years in conditions favourable for seed viability. The relative humidity is kept very low (15% in some cases), and the temperature is about +4°C for medium-term storage and about -18°C for long-term storage.

Infestation with insects (grain weevils, bruchids) is observed only occasionally in stored germplasm since the standard storage conditions are unfavourable for insect development. Insect infestation of seeds is easily detectable compared to infection or contamination with pathogens. However, germplasm is not in all cases stored under ideal conditions, and therefore insect pests also deserve attention.

Unfortunately, there are many examples where pathogens also survive even under the recommended storage conditions (Neergaard, 1977). Some pathogens survive for more than 10 years. Generally, one could expect at least a low percentage of the inoculum to survive as long as the seeds. Such infection may not alarm plant breeders and may be acceptable for seed production. However, to minimize the risk of spreading pathogens, a very low incidence of seed-borne inoculum warrants attention. When conditions are favourable, a disease can be transmitted even from traces of inoculum.

The health status of seeds is often related to the diversity of the stored germplasm. Diverse materials include lines which are resistant to specific pests or pathogens and others that may be susceptible. However, one cannot assume that the seeds of lines resistant to a pathogen, e.g. *Ascochyta* sp., are necessarily free from this pathogen. In fact, they might even carry extraordinarily virulent races. On the other hand, very susceptible lines can be free from the pathogen if grown under conditions unfavourable for disease development (in case of *Ascochyta*, a dry environment) or if the crop has been well protected by fungicide sprays during regeneration.

Another problem with germplasm and seed-borne pathogens is the maintenance of the genetic composition of the germplasm. Deliberate selection, i.e. the rouging of diseased plants in the field, is not recommended. Lines that are susceptible to a prevalent pest or pathogen could be eliminated altogether, although they may carry sources of resistance to other pests or pathogens. It is therefore important to control pests, diseases and weed during germplasm multiplication / regeneration. Curators are often reluctant to use pesticides, which might be phytotoxic to germplasm materials. To avoid the use of pesticides, hand-weeding and regeneration of germplasm at locations unfavourable for epidemics are procedures which are desirable and should be followed.

Collecting germplasm can result also in collecting pathogens, and with the movement of germplasm, there is a risk of spreading pathogens or their races to areas where they did not occur before. Undoubtedly in the centres of genetic diversity of germplasm there is also a diversity of pathogens and their races. From a quarantine point of view the question of races is very important, too. Many pathogens occur in different races, which are morphologically identical, but which differ in their virulence and host range. The problems with crop resistance breaking down upon introduction of new races are very real and can be catastrophic.

There are reports that germplasm can act as an effective vehicle for the dissemination of pathogens. Often plant introduction stations are faulted for the introduction and spread of plant pathogens. Hampton and Braverman (1979) gave an example of pea seed-borne mosaic virus on peas (*Pisum sativum*) being introduced to the Northeastern Regional Plant Introduction Station at Geneva, New York, USA. Kaiser and Hannan (1982, 1986) reported the incidence of *Ascochyta lentis* in lentil seeds imported to the United States. Bos (1977) discussed the risks of spreading viruses with germplasm collections.

Germplasm is generally obtained or collected from different geographical regions, and many accessions are multiplied together in a relatively small area at the research station or plant introduction station. If only one line is infected with an exotic pathogen, and conditions are favourable for disease development, it is likely that a large number of susceptible accessions will also become infected. Another reason for reports of exotic pathogens occurring at plant introduction stations could be that at least in some countries plant pathologists focus their attention on research stations rather than on farmers' fields and thus disease outbreaks are more readily detected in these stations.

In Syria an appreciable number of seed-transmitted diseases are endemic (Table 3). Some of these diseases are caused by pathogens that are considered quarantine organisms in countries that receive seeds from ICARDA, such as barley stripe mosaic virus, *Pseudomonas syringae* pv. *pisi*, *Xanthomonas campestris* pv. *translucens*, *Ditylenchus dipsaci*, *Ascochyta rabiei*.

At the same time, crops and genetic materials in Syria must be protected from the inadvertent introduction of new pests and diseases, such as Karnal bunt (*Tilletia indica*), dwarf bunt (*Tilletia controversa*), glume blotch (*Septoria nodorum*), and others, which are not endemic in this country. ICARDA is therefore taking appropriate precautions to avoid the dissemination of pests and pathogens.

Table 3. Some seed-borne diseases and pathogens which are endemic to Syria.

Disease	Pathogen	Host
Virus diseases	broad bean stain virus	faba bean
	bean yellow mosaic virus	faba bean
Bacterial blight	*Pseudomonas syringae* pv. *pisi*	pea
Leaf streak	*Xanthomonas campestris* pv. *translucens*	cereals
Loose smut	*Ustilago* spp.	cereals
Common bunt	*Tilletia* spp.	cereals
Ascochyta blight	*Ascochyta* spp.	legumes
Root and foot rot	*Fusarium* spp.	diff. crops
Stem nematode	*Ditylenchus dipsaci*	faba beans
Ear cockle	*Anguina tritici*	wheat

Activities of ICARDA's Seed Health Laboratory

The overriding objective of the Seed Health Laboratory of ICARDA's Genetic Resources Program is to dispatch and accept healthy seeds. The flow of germplasm is illustrated in Figure 1. All the incoming and outgoing seeds are inspected and tested by the Seed Health Laboratory. Phytosanitary certificates are only issued by the Syrian quarantine service after the necessary tests have been satisfactorily performed.

At ICARDA, incoming seeds are initially received at the Seed Health Laboratory for visual inspection. Then, depending on the country of origin, specific tests are conducted. Attention is given to all pests and pathogens, including those already occurring in Syria. With many of them, e.g. *Ascochyta rabiei* on chickpeas or *Tilletia caries* on wheat, there is still the risk of introducing new races. Special attention is given to pathogens that are known to occur in the introduced seeds' country of origin, especially when these pathogens have not yet been found in Syria. An example is *Tilletia indica*, a pathogen occurring in India, Pakistan and Mexico, causing Karnal bunt of wheat. Wheat from these countries is checked carefully for infected seeds. In addition, a centrifuge wash test, which reveals spore contamination at low levels, is conducted. When infected or contaminated seeds are detected, the material is not planted.

However, quite frequently information on the distribution of pests and pathogens is lacking or incomplete. Sometimes, there are no recommendations for suitable test methods. Therefore, all seeds are treated with broad-spectrum fungicides, such as benomyl + tridemorph + maneb for legumes and carboxin + thiram for cereals. If necessary, insects are controlled by fumigation with aluminiumphosphide or by cold treatment at -18°C. Then the seeds are handed over to scientists, who plant them in a designated 'isolation area' of the experiment station. During the vegetative growth of the crop, frequent field inspections are carried out to monitor the presence of diseases or pests. As yet, no quarantine glasshouse facilities are available, but plans are being considered to establish such facilities in the near future.

Also the fields where germplasm is multiplied for dispatch are inspected during the vegetative phase. At ICARDA all seed and germplasm multiplication is conducted at the main research station near Aleppo, 150 km from the coast in a dry climate. The coastal climatic conditions are favourable for the development of many pathogens, such as *Ascochyta* spp., *Ditylenchus dipsaci, Pseudomonas syringae* pv. *pisi*. These are used for screening germplasm for disease resistance. Also, intensive crop protection procedures are followed wherever appropriate. Faba beans, for example, are repeatedly treated with glyphosate to control *Orobanche* spp., fungicides (mancozeb, chlorothalonil and vinclozolin) to control *Ascochyta fabae* and *Botrytis fabae*, and insecticides (endosulfan or triazophos) to control *Bruchus* spp.

Later, random samples of harvested seeds are tested. Special attention is given to samples from fields where potentially seed-borne diseases were found during field inspection. Laboratory tests, for example the blotter test and the agar test (Neergaard, 1977), which are suitable for a broad spectrum of pathogens, are routinely used. If a recipient country has specified certain pathogens or pests as quarantine organisms, appropriate tests will be conducted wherever possible to ensure freedom from these pathogens or pests. Many European countries, for example, request an additional declaration on the phytosanitary certificate proclaiming freedom from *Pseudomonas syringae* pv. *pisi* in pea seeds. Therefore, all pea seeds dispatched to those countries are tested on a differential medium (King's B medium); accessions showing a positive reaction are not sent. Other countries, for example Turkey, request wheat seeds which are free from *Urocystis tritici* (flag smut) spores. Representative samples of wheat seeds are therefore tested by the centrifuge wash test (Neergaard, 1977) before dispatch, even if the disease has not been observed on the crop during field inspection. The laboratory tests are complementary to field inspection. A very low disease incidence not observed in the field, would be detected by a sensitive test procedure, e.g. the centrifuge wash test.

All seeds are treated before dispatch from ICARDA, unless specific requests are made for untreated seeds. Efforts are made to select safe fungicides that have a wide spectrum of activity without having phytotoxic effects. Systemic fungicides, such as carboxin, benlate, etc. are preferred. Legume seeds, which could be infested by various storage pests, such as bruchids, are routinely fumigated with aluminiumphosphide.

Facilities for routine testing of seeds for different legume viruses and barley stripe mosaic virus by ELISA have been developed.

Testing and treatment of seeds

The general principles of seed health testing and treatment of commercial seed lots are of course applicable also for germplasm material. However, with germplasm one should be aware of a different set of problems. Some of these are:

Sampling

For commercial seed there are strict rules to be followed in order to obtain a representative sample. However, the 'seed lot' of germplasm is frequently limited to just a few grams. Adequate sampling is therefore difficult since only a very limited number of seeds is available for testing. This naturally reflects on the reliability of the tests, as the results obtained may not be reproducible.

Testing

With the small number of seeds available for testing, it is extremely difficult to obtain enough data. Usually several different tests, such as blotter test, centrifuge wash test, agar test, tests for viruses, etc. should be carried out in order to identify as many different pathogens as possible. The test methods used are mostly those described by Neergaard (1977) and the ISTA Handbook of Seed Health Testing, Section 2 (1977-1986). Unfortunately most of the methods destroy the seeds. The testing of soil, plant debris, broken seeds, etc. that are removed during seed cleaning can indicate roughly what pathogens might be present in the seed lot, but does not provide information on the incidence of infection.

Seed recovery after testing

Because of the limited number of seeds for some entries, it is often desirable to recover viable seeds following testing. In some cases, seedlings from the blotter test are transplanted and grown for reproduction. Seeds from the centrifuge wash test can be redried with only slight loss in viability, provided water and not an organic solvent has been used to remove seed treatment chemicals.

Another way is to make use of standard germination tests that are carried out regularly to assess seed viability and to determine the time for regeneration. However, this method limits the testing only to certain pathogen species. Important pathogens such as *Tilletia* spp., *Ditylenchus dipsaci* and many others cannot be detected by this method.

Field inspection

Meticulous field inspection is gaining importance because of these problems associated with seed health testing of germplasm. For seed certification field inspection for diseases is a standard requirement. If a seed production field of several hectares planted with one variety is properly sampled and carefully inspected, one can obtain reliable information on the disease situation in the field. With germplasm multiplication the situation is different. Thousands of lines which differ greatly in their susceptibility to pests and pathogens are grown in a field the size of a normal seed multiplication field (about 10 hectares). Sample areas for inspection are therefore inadequate. Each line must be inspected separately. This is a time-consuming task, which must complement the seed health testing to obtain additional information on the health status of seeds. Another important problem encountered with field monitoring of germplasm is the difficulty in detecting certain pathogens. For example, Karnal bunt (*Tilletia indica*) is not easily diagnosed in the field; some viruses may also be latent, i.e. the host does not show any symptoms. Generally, it is difficult to detect diseases that occur at a low incidence.

Seed treatment

For commercial seed production a wide range of seed treatment equipment is available. Some of the machines are suitable for liquid or slurry and for powder dressing; they are usually equipped with automatic dosage devices, which ensure exact dosage and equal distribution of the chemicals on the seeds. The capacity of these seed treaters varies from 0.5 ton to more than 20 tons of seeds per hour.

Only a limited range of equipment is available for the treatment of 'seed lots' ranging from few grams to several kilograms. Such machines are designed almost exclusively for research purposes, and are not suitable for continuous operation, that is, for treatment of a large number of small samples within a relatively short time. Mixing of the seeds from different entries must be avoided.

Usually, liquid or slurry treatment, which is less hazardous for the operator, is preferred to dust treatment. Liquid or slurry treatment gives a better coverage of the seeds, and the required dosage can be applied more precisely. At ICARDA a simple machine suitable for the application of liquid and slurry is being used. Seed samples of 20 g to 4 kg can be treated. The machine consists of a stainless steel bowl that comes in three sizes, a motor that ensures even mixing, a container with an automatic dosage pipette for the treatment chemicals, and a frame. This piece of equipment is commercially available from Messrs. Hege, Hohebuch, 7112 Waldenburg, West Germany.

The normal slurry treatment, although superior to dust treatment, may not be suitable for the eradication of some pathogens. At ICARDA it was found that soaking the seeds in fungicide or antibiotic solution is more effective than conventional treatment. Soaking pea seeds for one hour in a 500 ppm streptomycin solution is very effective in controlling *Pseudomonas syringae* pv. *pisi* (Diekmann, 1984). The seeds must be redried carefully in order to maintain seed viability. This procedure, although time-consuming, is currently being used to free valuable germplasm from *Pseudomonas syringae* pv. *pisi* before dispatch to countries where this pathogen is considered a quarantine object.

Conclusion

The exchange of genetic material is necessary for crop improvement. However, it entails the risk of spreading pests and pathogens or their races to unaffected areas. This risk can be minimized by applying seed health measures, such as (1) field inspection for pathogens or pests that are seed-transmitted, (2) laboratory seed health testing, and (3) seed treatment. A combination of these measures would effectively reduce the risk of pathogen dissemination. It should be acknowledged that there cannot be a 100% safety against the introduction of pathogens. In order to further control the spread of pests and pathogens with germplasm, increased cooperation between quarantine services and suppliers of germplasm is desirable. Lists of seed-borne pests and pathogens in any particular country should be available for use by quarantine services. The use of a Plant Germplasm Health Certificate as suggested by the FAO/IBPGR Task Force on safe germplasm transfer (Hewitt and Chiarappa, 1977) would provide additional information on the results of tests and on any other measures, e.g. field inspections or multiplication by meristem culture, which have been conducted in the country of origin.

The risk of introducing pests and pathogens or their races must be seriously considered along with the benefit of broadening the genetic base of crop plants. For the assessment of this risk it should be realized that mere dissemination of a pathogen does not necessarily mean transmission of a disease. An outbreak of a disease requires successful infection, which depends largely on favourable environmental conditions. Since the environmental conditions in recipient countries are not always known, ICARDA has adopted reasonable and effective precautionary measures to ensure that pathogen-tested germplasm is distributed from, and accepted at its research station.

Acknowledgements

I wish to thank Dr. B.H. Somaroo, Head of ICARDA's Genetic Resources Program, for his helpful comments and valuable suggestions for improving this manuscript.

References

Bos, L. 1977. Seed-borne viruses. pp. 39-69. In: *Plant Health and Quarantine in International Transfer of Genetic Resources*. (eds. W.B. Hewitt and L. Chiarappa). CRC Press, Cleveland, Ohio.

CGIAR 1980. Consultative Group on International Agricultural Research. CGIAR Secretariat, Washington.

Diekmann, M. 1984. Saatgutbehandlung von Koernerleguminosen - Problematik und Bedeutung im Internationalen Saatgutaustausch [Seed treatment of grain legumes - problems and importance for international seed exchange] (abstr.). *Mitteilungen der Biologischen Bundesanstalt für Land- und Forstwirtschaft H.* **223:** 140.

Hampton, R.O. and Braverman, S.W. 1979. Occurrence of pea seed-borne mosaic virus and new virus-immune germplasm in the plant introduction collection of *Pisum sativum*. *Plant Disease Reporter* **63**: 95-99.

Hewitt, W.B. and Chiarappa, L. (eds.) 1977. *Plant Health and Quarantine in International Transfer of Genetic Resources*. CRC Press, Cleveland, Ohio.

International Seed Testing Association. 1977-1986. *Handbook of Seed Health Testing*. Section 2, Working Sheets, Published at irregular intervals. ISTA, Zurich, Switzerland.

Kaiser, W.J. and Hannan, R.M. 1982. *Ascochyta lentis*: incidence and transmission in imported lentil seed (abstr.) *Phytopathology* **72:** 944.

Kaiser, W.J. and Hannan, R.M. 1986. Incidence of seedborne *Ascochyta lentis* in lentil germ plasm. *Phytopathology* **76**: 355-360.

Neergaard, P. 1977. *Seed Pathology*. Vol. I & II. The Macmillan Press Ltd., London. 1191 pp.

Richardson, M.J. 1979. *An Annotated List of Seed-borne Diseases*. 3rd edition. International Seed Testing Association, Zurich, Switzerland.

Richardson, M.J. 1981. *Supplement 1 to an Annotated List of Seed-borne Diseases*. 3rd edition. International Seed Testing Association, Zurich, Switzerland.

Richardson, M.J. 1983. *Supplement 2 to an Annotated List of Seed-borne Diseases*. 3rd edition. International Seed Testing Association, Zurich, Switzerland.

Vavilov, N.I. 1951. Phytogeographic basis of plant breeding. In: *The Origin, Variation, Immunity and Breeding of Cultivated Plants*. Translated by K. Starrchester. Ronald Press Co., New York, USA.

Germplasm Exchange and Plant Quarantine Systems at ICRISAT

MELAK, H. MENGESHA and N.C. JOSHI

International Crops Research Institute for the Semi-Arid Tropics, Patancheru, Andhra Pradesh 502 324, India

Introduction

The germplasm is considered as a part of human biological heritage without whose free exchange and availability, present day farm productivity would not have been possible (Jain, 1982). Therefore, collection, evaluation, utilization, conservation and exchange of genetic resources assume considerable significance, especially in view of the rapid degradation and exploitation of the available biodiversity all over the world (Mehra and Arora, 1982; Mengesha, 1984; Paroda and Arora, 1986; Paroda, 1989). Considerable efforts are being made worldwide to conserve the genetic resources of important crop plants. Free exchange of diverse germplasm is essential. Genetic manipulation is advancing at a fast pace and the progress may even be more accelerated by the application of genetic engineering techniques in the future (Law, 1986).

The role played by the Consultative Group on International Agricultural Research (CGIAR) in the collection and conservation of germplasm is commendable. Through the catalytic role and direction action of the International Board for Plant Genetic Resources (IBPGR), endangered germplasm of many crops has been collected and conserved in many gene banks (IBPGR, 1990). The International Agricultural Research Centers (IARCs) strategically located in regions of rich crop diversity are in a unique position to collect, conserve, and evaluate germplasm and make it readily available to all scientists throughout the world. Several national programmes have also made substantial efforts to collect and conserve their indigenous plant genetic resources. Although enormous genetic wealth is being preserved by national, regional, and international organizations, its potential and usefulness largely depends on its health, viability, and free exchange among the users of the material.

In India, the National Bureau of Plant Genetic Resources (NBPGR) is the primary organization which is responsible, *inter alia*, for the exchange of germplasm of agri-horticultural crops between India and other countries (Joshi *et al.*, 1989). NBPGR maintains exchange links with about 70 countries including IARCs.

Germplasm exchange at ICRISAT

In the area of genetic resources, ICRISAT's mandate crops include: sorghum (*Sorghum bicolor*), pearl millet (*Pennisetum glaucum*), pigeonpea (*Cajanus cajan*), chickpea (*Cicer arietinum*), groundnut (*Arachis hypogaea*) along with six minor millets: finger millet (*Eleusine coracana*), foxtail millet (*Setaria italica*), proso millet (*Panicum miliaceum*), little millet (*Panicum sumatrense*), barnyard millet (*Echinochloa crusgalli*), and kodo millet (*Paspalum scrobiculatum*).

The major activities of ICRISAT's Genetic Resources Unit are to:

(i) collect, assemble, and conserve the germplasm of its mandate crops and their wild relatives;
(ii) characterize, evaluate, and document the germplasm;
(iii) maintain and rejuvenate the germplasm without altering the original genotype or population; and
(iv) distribute and exchange healthy germplasm for present and future utilization.

The numbers of germplasm accessions assembled by ICRISAT from various countries and samples distributed to 147 countries are summarized in Table 1. For each of the ICRISAT mandate crops, these accessions represent the largest collection of germplasm assembled at any one place. However, considering the extent of the area devoted to these crops worldwide, the present collections are still too small.

Table 1. Details of germplasm assembly and distribution from Genetic Resources Unit, ICRISAT

	Assembly		Distribution			
Crop	Accessions	No. of Countries	ICRISAT*	India	Abroad	No. of Countries
Sorghum	31 817	87	215 649	86 396	108 855	93
Pearl millet	21 772	44	24 827	44 852	29 540	69
Pigeonpea	11 482	54	57 058	30 011	12 929	97
Chickpea	15 941	42	98 358	41 202	45 765	76
Groundnut	12 712	89	39 890	30 616	25 693	83
Minor millets	6 610	24	-	17 901	12 294	33
Total	100 334		435 782	250 978	235 076	

* For use in ICRISAT crop improvement programme

ICRISAT is serving as a world depository of genetic resources of its mandate crops. Several thousand accessions of the mandate crops including the six minor millets are conserved in medium-term (+4oC and about 20% relative humidity) cold storage facilities for exchange purposes (Mengesha, 1984). About 500 g seed of each accession is dried to about 5-8% moisture content before it is stored. The type of storage chambers used at ICRISAT and the general standards and system of germplasm technology and conservation have been described elsewhere (Mengesha *et al.*, 1989). Long-term cold storage (-20oC) chambers have also been installed and are now operational as a part of ICRISAT gene bank. Germination tests conducted in September 1990 on conserved seeds showed over 92% viability after nine years of storage. All the seeds are rejuvenated before their germination drops below 85%, or before the seed quantity reaches a critical low level.

So far 100,334 accessions of various ICRISAT crops are being conserved (Mengesha, 1988). A small portion of the conserved germplasm has so far been effectively utilized in various plant breeding programmes especially as donors of certain desirable traits like resistance to biotic and abiotic stress factors. Much broader application and impacts are envisaged with the realization and imaginative manipulation of the yet unknown and potentially useful genetic traits of the majority of the conserved and introduced germplasm.

Role of quarantine in exchange of germplasm at ICRISAT

The success of international germplasm exchange and utilization largely depends on its timely transfer and ease of mobility. Unfortunately, however, there are many hurdles that a germplasm sample has to pass through before it can reach its destination. From the plant quarantine point of view, it is true that there is some risk in the transfer of unchecked germplasm from one region to another (Kahn, 1977). Likewise, it may be stated that the aim of germplasm collection and exchange is to conserve and introduce useful germplasm without endangering the new habitat. Therefore, there is something in common between quarantine and genetic resources - both are necessary and useful.

While exchanging germplasm for utilization in crop improvement programmes, the samples have to pass through the National Plant Quarantine system which ensures that only healthy seeds are exported. A safe and rapid transfer of germplasm is vital for a sound crop improvement programme. A great many exotic crops are flourishing in many areas of the world as a result of international transfer and exchange of germplasm. Import of small, experimental quantities of seeds with appropriate safeguards based on sound biological principles can often be an answer to improve the genetic base of crops. Much larger quantities of commercial seed often enter a country with even greater quarantine risk but often with minimal or cursory inspection. Yet, germplasm is the basic raw material for future development of commercial seed and to overimpose undue restrictions on its movement appears to be counter-productive.

Quarantine system in India

In India, the import and export of plants and plant materials are regulated by the rules and regulations framed under Destructive Insects and Pests (DIP) Act of 1914. Subsequently the Act was revised eight times by the Government of India (GOI). The main objective of the Act is to prevent the introduction into, and the transport from one state to another within India of any insect, fungus, and other pest which is or may be destructive to crops (DPPQS, 1976; Joshi, 1975; Joshi *et al.*, 1989).

Originally, all seeds were not included in the DIP Act. Later in 1984 GOI passed the Plants, Fruits, and Seeds Order, which came into effect in 1985. Under this order 17 crops are included, and the conditions for their import are stipulated. The main features of this order are as follows:

1. To bring seed under the purview of DIP Act.
2. To regulate importation of seeds only through valid import permit issued either by the Plant Protection Adviser to the GOI or by the Director, NBPGR, in respect of import of seed and plant material for research purpose made by any institute of the Indian Council of Agricultural Research, Agricultural Universities, and ICRISAT (DPAC, 1985; DPAC, 1987; NBPGR, 1987).
3. To permit entry of seed consignments only if they are accompanied by an official phytosanitary certificate (PSC) issued by the quarantine authority of the exporting country.
4. Stipulating post-entry isolation growing of specified crops at the approved locations.
5. No consignment wherein hay or straw or any material of plant origin used for packing shall be imported.
6. Import of soil earth, compost, sand, plant debris, etc. along with plants, seeds shall not be permitted. The NBPGR is now the authorized agency in India that controls, serves, and regulates the plant quarantine requirements of ICRISAT.

Quarantine system at ICRISAT

GOI has authorized ICRISAT unrestricted movement of seeds and genetic material of its mandate crops into and out of India as required for collaborative work in any part of the world consistent with the appropriate quarantine regulations prevailing in the country.

GOI took a number of steps to facilitate smooth clearance of germplasm as well as to check introduction of exotic pests or diseases of quarantine importance. In 1973, it declared the Central Plant Protection Training Institute (CPPTI), Rajendranagar, Hyderabad, as the quarantine authority to clear ICRISAT's seed material of its mandate crops. In 1978, ICRISAT was permitted to set up an Export

Certification Quarantine Laboratory at its campus under the overall authority of CPPTI. Recently in 1986, the GOI authorized NBPGR as the sole plant quarantine authority to clear ICRISAT's mandate crops as well (NBPGR, 1986). NBPGR has set up a Regional Station at Hyderabad, which carries out the quarantine clearance for ICRISAT material. Excellent collaboration exists between ICRISAT and NBPGR.

Table 2. Seed treatment schedules for quarantine clearance of ICRISAT mandate crops.

Crop	Treatment	Remarks (if any)
Sorghum	a. Thiram or captan at 2.5 g kg^{-1}	Prophylactic measure against seed rot or moulds
	b. Carboxin at 1.5 g kg^{-1}	Against smut disease
	c. Metalaxyl at 4 g kg^{-1}	Against downy mildew
Chickpea	a. Mixture of benomyl and thiram (3:2) at 4.5 g kg^{-1}	Against wilt disease
	b. Thiabendazole at 3 g kg^{-1}	Against Ascochyta blight
Pigeonpea	Mixture of benomyl and thiram (3:2) at 4.5 g kg^{-1}	Against wilt diseases
Groundnut	Thiram at 3 g kg^{-1} of seed	Against fungal pathogens
Pearl millet	a. The seeds are soaked in 0.1% mercuric chloride for 10 min and then thoroughly washed in running water for 5 min.	Against downy mildew
	b. After washing, the seeds are transferred immediately into a water bath set at 55°C for 12 min.	
	c. Immediately after the hot water treatment, the seeds are transferred to water at room temperature for 2 min cooling after which it is transferred to incubator at 35°C for 12 h and then at 40°C for an additional 12 h.	
	d. After cooling the seeds are treated with the fungicide metalaxyl (Ridomil 50% WP) at 3 g kg^{-1} seed in 1000 ml of 1% aqueous methyl cellulose solution. The seeds are soaked in the fungicide suspension for 4-6 h. The treated seeds are dried under shade/ sunlight and can be used for sowing up to a period of 4 months.	

Quarantine procedures for import of germplasm

All the germplasm or seed certified by the national quarantine services of the exporting country, and accompanied by a phytosanitary certificate, is received at NBPGR. It is fumigated at normal atmospheric pressure with aluminium phosphide (dosage: 3 g m^{-n} for 5 days). The seeds are then closely examined for foreign matter including smut sori, ergot sclerotia, weed seeds, nematode cysts, and soil clumps. The seeds are then subjected to various seed health testing procedures like washing and sedimentation test, blotter/agar plate tests, and X-ray radiography for detection of latent infestation (Nirula, 1979; Pathak and Joshi, 1984). The seed samples of sorghum, pearl millet, chickpea, pigeonpea, and minor millets are then treated with approved pesticides before release. The treatment schedule followed for ICRISAT mandate crops is given in Table 2 (Varma and Ravi, 1984).

Groundnut is grown for six weeks in an insect-proof screenhouse and thoroughly checked for symptoms of the following exotic virus diseases: peanut mottle, peanut stunt, marginal chlorosis, peanut stripe, and ring spot. Healthy seedlings are then released for planting in the post-entry quarantine isolation area (PEQIA) at ICRISAT centre. For the last two years, enzyme-linked immunosorbent assay (ELISA) test has been used in ICRISAT for detection of viruses (Bharathan *et al.*, 1984; Prasada *et al.*, 1988).

The cuttings of wild species of *Arachis* originating from South and North America are subjected to intermediate plant quarantine in a non-groundnut growing country. In this regard there is a long-standing agreement between ICRISAT and the University of Reading in UK to grow and examine groundnut cutting specimens at Reading before the healthy plants are airlifted to India for further quarantine examination and final entry into the country. The material is then released to ICRISAT for planting in quarantine nethouse, where plants are grown, well established, and thoroughly checked before they are transplanted in PEQIA.

Additional requirements for import

As per the particular national quarantine regulations, specific additional declarations are required to be stated in the PSC as a safeguard against specific pests and diseases whose introduction can be considered as a high risk to the mandate crops in India. The requirements are as follows:

Sorghum: Seed samples to be certified as collected from fields free from bacterial leaf stripe (*Pseudomonas andropogonis*), bacterial leaf streak (*Xanthomonas campestris* pv. *holcicola*), and southern leaf blight (*Drechslera maydis*).

Pearl millet:	Seeds should be certified as collected from crop free from downy mildew infection (*Sclerospora graminicola*).
Chickpea:	Seed samples to be certified as collected from plants free from blight (*Ascochyta rabiae*) and virus diseases.
Pigeonpea:	No specific requirement.
Groundnut:	Seeds to be certified as collected from fields free from peanut mottle, peanut stunt, marginal chlorosis, peanut stripe virus, rust (*Puccinia arachidis*), and scab (*Sphaceloma arachidis*).

Postentry quarantine isolation area

Postentry Quarantine Isolation Area (PEQIA) is about 50 ha located in the southeast corner of ICRISAT farm. About 6 ha of it is under cropping. This area is 200 m away from the nearest crops fields. All seed materials released by NBPGR except chickpea are required to be grown for one generation in the PEQIA at ICRISAT Center. The crops are raised under close, joint supervision of the staff of NBPGR and ICRISAT. They inspect crops biweekly from sowing to harvest. Plants showing symptoms of exotic diseases are promptly rogued and incinerated. Seeds harvested from healthy plants are released to ICRISAT scientists.

During 1973-89, 77,164 seed samples of sorghum, 25,140 of pearl millet, 28,861 of chickpea, 8,716 of pigeonpea, 14,870 of groundnut, 6,174 of minor millets, totalling 160,925, and 5,000 groundnut cuttings, were imported by ICRISAT from 95 countries.

Interceptions

The national plant quarantine authorities have intercepted some important insects and pathogens of plant quarantine significance. A list of the interceptions made during 1975-1987 is given in Table 3 (Wadhi, 1980; Pathak and Joshi, 1984; NBPGR, 1987).

Quarantine procedures for seed export

The quarantine regulations for the export of seed material are based on the 1951 International Plant Protection Convention and are modified from time to time according to specific requirement of the importing country. However, it is obligatory for the seed consignors in all countries to ensure that the seed is examined and cleared by the concerned national plant quarantine authorities and is accompanied by a PSC. The seed export from ICRISAT started in 1974. In 1978, GOI approved the establishment of an Export Certification Quarantine Laboratory at ICRISAT with all infrastructure facilities i.e. vacuum, and atmospheric fumigation chambers, radiographic equipment, incubation room, and rooms for inspection, seed treatment, packeting, cold store, and well qualified staff for the expeditious movement of germplasm to other countries under the overall authority of the national plant quarantine service.

The procedures for seed inspection and fumigation treatments are similar to those for imported material prescribed by the national plant quarantine agency.

Phytosanitary requirements of the importing countries

Most germplasm importing countries have their own specific quarantine rules and regulations which ICRISAT respects and follows strictly. The quarantine rules and regulations of all countries that are available at ICRISAT are used as specific guidelines for germplasm exchanges. The rules and regulations are updated from time to time through information obtained from the Food and Agriculture Organization of the United Nations (FAO), European and Mediterranean Plant Protection Organization (EPPO), United States Department of Agriculture (USDA), and other agencies.

Table 3. Pests intercepted by NBPGR from plant materials imported during 1975-1987.

Pests	Host	Country
Insect pests		
Acanthoscelides obtectus	*Cajanus cajan*	Brazil
Specularius erythraeus	*Cajanus cajan*	Brazil
Bruchidius sp.	*Cajanus cajan*	Senegal
Callosobruchus analis	*Cajanus cajan*	Thailand
Callosobruchus chinensis	*Cajanus cajan*	Afghanistan
Callosobruchus maculans	*Cajanus cajan*	Iraq
Corcyra cephalonica	*Arachis hypogaea*	USA
Ephestia cautella	*Arachis hypogaea*	Canada
Oryzaephilus surianamensis	*Pennisetum glaucum*	Sudan
Gonocephalum sp.	*Sorghum bicolor*	USA
Sitophilus ziamais	*Sorghum bicolor*	Zimbabwe
Pathogens		
Drechslera maydis	*Sorghum bicolor*	USA
Drechslera sorghicola	*Sorghum* sp.	Australia
Fusarium moniliforme	*Sorghum* sp.	Australia
Perenosclerospora sorghi	*Sorghum* sp.	Zimbabwe
Colletotrichum graminicola	*Sorghum* sp.	Italy
Fusarium solani	*Sorghum* sp.	Kenya
Xanthomonas campestris pv. *holcicola*	*Sorghum* sp.	Yemen Arab Rep.
Pseudomonas andropogoni	*Sorghum* sp.	Yemen Arab Rep.
Gloeocercospora sorghi	*Sorghum bicolor*	Italy
Drechslera sorghicola	*Pennisetum glaucum*	Mali
Drechslera maydis	*Pennisetum glaucum*	USA
Tolyposporium penicillariae	*Pennisetum glaucum*	Italy
Claviceps sp.	*Pennisetum glaucum*	Zimbabwe
Fusarium solani	*Pennisetum glaucum*	USA
Nematodes		
Ditylenchus angustus (?)	*Arachis hypogaea*	USA
Aphelenchoides besseyi	*Setaria italica*	Bangladesh

Total exports from ICRISAT

During 1974-1989, 368,980 germplasm and breeders' seed samples of sorghum, 131,326 of pearl millet, 184,548 of chickpea, 48,162 of pigeonpea, 59,357 of groundnut, and 12,047 of minor millets, totalling 804,420 were exported from ICRISAT to 147 countries. So far, there has been no report of introduction of any pest or disease through exchange of ICRISAT germplasm.

Future perspectives and suggestions

For safe exchange of germplasm we suggest:

a) More extensive use of enzyme-linked immunosorbent assay (ELISA) technique in the detection of viruses and bacterial pathogens.
b) Exploration of tissue culture technique wherever possible in exchange of germplasm.
c) Adoption of new disease indexing techniques.
d) Updating the treatment schedules, procedures under different environmental conditions.
e) Compilation of information on pests, diseases, and weeds of plant quarantine importance and outbreaks on a regional and global basis.
f) Avoidance of bulk import of seeds to facilitate thorough inspection.
g) Development of collaborative research work among genetic resource, pathology, entomology and plant quarantine scientists on seed-borne diseases, seed health, and treatment schedules of ICRISAT mandate crops.
h) Joint preparation of a plant quarantine manual by ICRISAT and NARS for the guidance of all countries involved in germplasm exchange.
i) Safeguard against introduction of exotic pests and pathogens. Germplasm of some crops may require closer observation for one or more seasons, hence well equipped glass or screen houses for quarantine isolation and screening of such material are required.
j) Create awareness among the public and planners about the importance of germplasm and use of plant quarantine for international germplasm exchange for higher and sustainable crop production.

References

Bharathan, N., Reddy, D.V.R., Rejeshwari, R., Murthy, V.K. and Rao, V.R. 1984. Screening peanut germplasm lines by enzyme-linked immuno-sorbent assay for seed transmission of peanut mottle virus. *Plant Diseases* **68**: 757-758.

DPAC (Department of Agriculture and Cooperation) 1985. *The Plants, Fruits, and Seeds (Regulation of Imports into India) Order, 1984, 1-25*. Ministry of Agriculture, New Delhi, India.

DPAC (Department of Agriculture and Cooperation) 1987. *The Plants, Fruits, and Seeds (Regulation of Imports into India) Order, 1987, 1-3*. Ministry of Agriculture, New Delhi, India.

DPPQS (Directorate of Plant Protection, Quarantine & Storage) 1976. *The Destructive Insects and Pests Act, 1914 (2 of 1914) and the rules made there under upto June 1976, 1-67*. Ministry of Agriculture, New Delhi, India.

IBPGR (International Board for Plant Genetic Resources) 1990. *Annual Report 1989*. IBPGR, Rome.

Jain, H.K. 1982. Plant breeders' rights and genetic resources. *Indian Journal of Genetics and Plant Breeding* **42**: 121-128.

Joshi, N.C. 1975. Regulatory measures in controlling plant diseases in India. *Pesticides* **10**: 18-21.

Joshi, N.C., Singh, B.P., Ram Nath and Varaprasad, K.S. 1989. Germplasm exchange and quarantine in India. pp. 75-78. In: *Collaboration on Genetic Resources: Summary of the Proceedings of a Joint ICRISAT/NBPGR (ICAR) Workshop on Germplasm Exploration and Evaluation in India*. November 14-15, 1988, ICRISAT Center, Patancheru, India.

Kahn, R.P. 1977. Plant quarantine: principles, methodology, and suggested approaches. pp. 289-308. In: *Plant Health and Quarantine in International Transfer of Genetic Resources* (eds. W.B. Hewitt and Chiarappa) CRS Press, Boca Raton.

Law, C.N. 1986. The need for a multidisciplinary approach to genetic manipulation in plant breeding. pp. 867-883. In: *Genetic Manipulation in Plant Breedings* (eds. Horn, Jensen, Odenbach, and Schieder). Breeding: Proceedings international symposium Sept. 8-13, 1985, Berlin, EUCARPIA, Walter de Gruyter, Berlin/New York.

Mehra, K.L. and Arora, R.K. 1982. Plant genetic resources of India - their diversity and conservation. *NBPGR Scientific Monograph* **4**: 60.

Mengesha, M.H. 1984. International germplasm collection, conservation, and exchange at ICRISAT. pp. 47-54. In: *Conservation of Crop Germplasm - a International Perspective*. Crop Science Society of America, Madison, USA.

Mengesha, M.H. 1988. Genetic resources activities at ICRISAT. *Indian Journal of Plant Genetic Resources* **1**: 49-58.

Mengesha, M.H., Khanna, P.P., Chandel, K.P.S. and Kameswara Rao. 1989. Conservation of world germplasm collection of ICRISAT mandate crops. pp. 65-70. In: *Collaboration on Genetic Resources: Summary of the Proceedings of a joint ICRISAT/NBPGR (ICAR) Workshop on Germplasm Exploration and Evaluation in India*. Nov. 14-15, 1988, ICRISAT, Patancheru, India.

NBPGR (National Bureau of Plant Genetic Resources) 1986. *Guidelines for the Exchange of Seed Plant Material*. pp. 1-5.

NBPGR (National Bureau of Plant Genetic Resources). 1987. *Annual Report 1987*. New Delhi, India: NBPGR. 188 pp.

Nirula, K.K. 1979. Quarantine regulations on seed imports at International Crops Research Instituteof the Semi-Arid Tropics. *FAO Plant Protection Bulletin* **27**: 119-122.

Paroda, R.S. 1989. Keynote address: Collaborative efforts on plant genetic resources by ICRISAT and ICAR. pp. 7-10. In: *Collaboration on Genetic Resources: Summary of the Proceedings of a Joint ICRISAT/NBPGR (ICAR) Workshop on Germplasm Exploration and Evaluation in India*. November 14-15, 1988, ICRISAT Center, Patancheru, India.

Paroda, R.S. and Arora, R.K. 1986. Plant genetic resources - an Indian perspective. *NBPGR Scientific Monograph* **10**: 34.

Pathak, V.K. and Joshi, N.C. 1984. Screening of germplasm material meant for ICRISAT at Central Plant Protection Training Institute. *Plant Protection Bulletin* **36**: 45-49.

Prasada, R.D., Rao, V.J. and Chakrabarty, S.K. 1988. Quarantine procedures for seed-borne virus of groundnuts. Paper presented at the *National Symposium on Plant virus problems in India*. April 5-6, 1988, Indian Agricultural Research Institute, New Delhi, India.

Varma, B.K. and Ravi, U. 1984. Plant quarantine facilities developed at ICRISAT for export of germplasm. *Plant Protection Bulletin* **36**: 37-43.

Wadhi, S.R. 1980. Plant quarantine activity at the National Bureau of Plant Genetic Resources. *NBPGR Scientific Monograph* **2**: 99.

Seed Movement and Quarantine Measures in Selected Near East Countries

A.M. ABDELMONEM

Seed Pathology Research Department, Plant Pathology Research Institute, Agricultural Research Center, Giza, Egypt

Introduction

Over the last few years, seed movement and germplasm exchange has increased spectacularly in international trade and agricultural development.

Size of individual seed consignments vary from shiploads of food grains, or tons of planting seed, down to a few grams for research or gardens (Neergaard, 1980). At the same time, accelerated worldwide movement of genetic material for breeding and scientific use have considerably increased the international spread of crop pests, diseases and weeds. Increased international movement of seed and germplasm constitutes real hazards for crop production (Karpati, 1981, 1983). Numerous destructive pests and diseases have been spread, with disastrous losses in hitherto uninfested areas, because effective quarantine precautions were not taken (Reddy, 1970; Neergaard, 1979).

Needed in international movement of seed is a special policy which requires specific essential provisions.

This paper describes current seed movement and quarantine measures in selected Near East countries, with emphasis on Egypt.

Interception and introduction of seed-borne diseases

Some examples of successful interception of seed-borne pathogens in seed movement in the Near East serve to illustrate the importance of quarantine measures.

Leppik (1962) reported that downy mildew (*Plasmopara halstedii*) of sunflower (*Helianthus annuus*) evidently a new pathogenic race from eastern Europe, was found on plants grown from seed received from Turkey and Pakistan.

A destructive leaf spot (*Cercospora traversiava*) on fenugreek *(Trigonella foenum-graecum)*, known in some eastern European countries and now common in the Near East, was found on seed samples from the Near East (Neergaard, 1979).

In 1952, USA intercepted *Tilletia pancicii* on *Hordeum* sp. from Turkey, and *Neovossia indica* on wheat seed from Afghanistan (Locke and Watson, 1955).

Post-entry inspection revealed a number of seed-borne diseases, including pathogenic races of three species of fungi, all new to the USA, including two from Pakistan and two from Turkey (Neergaard, 1979).

Leppik (1964) reported that squash mosaic virus was introduced to the USA by seed from Iran, and disseminated in Iowa by cucumber beetles. After several years of intensive work, the disease was eradicated and disease-free seed produced. Abdelmonem *et al.* (1989) reported that the blast epidemic in 1984 was due to introduction of specific races of *Pyricularia oryzae* with rice seed of cultivar Reiho, introduced from Japan which was released for commercial production in 1984.

Tobacco blue mould (*Peronospora tabacina*) was endemic in Australia and America until 1958, when it was reported in England. In a few years, it spread to all tobacco-growing areas of Europe and parts of North Africa and the Near East (Karpati, 1983).

Heavy chickpea losses in India in 1981 were due to introduction of a virulent biotype of *Ascochyta rabiei* from the Middle East.

Safe movement and quarantine legislation

Egypt exports berseem clover seed (*Trifolium alexandrium*) to Pakistan and Saudi Arabia, and seedlings of ornamentals and fruits to Saudi Arabia, Emirates, Oman, Qatar, Bahrain, Libya and the Arabian Islands.

Plant quarantine laws and regulations are based on national and international legal obligations. They have evolved in various parts of the world over the last century.

Following the International Plant Protection Convention established by FAO on December 6, 1951, Egypt's Law 53 of 1966 provides guidelines and safeguards with respect to plant introduction, import and export quarantine procedures. Egyptian phytosanitary legislation is regulated by a series of orders issued under this law. They include lists of prohibited plant pests and diseases, prohibited plants, provisions for releasing consignments for export or import, and many other aspects under the jurisdiction of the Ministry of Agriculture (Anonymous, 1967). These regulations are amended as conditions require.

Classification of plants for introduction into Egypt

Introduction prohibited. Importation of the following plants are prohibited but can be made for scientific purposes by official agencies with the approval of the Ministry of Agriculture. These are: *Gossypium* spp., *Hibiscus* spp. except dry roselle (*Hibiscus sabdariffa*) and Kenaf (*Hibiscus cannabinus*), *Abutilon* spp., *Althea* spp., *Corchorum officinarum*, *Vitis vinifera* (except fruit), *Citrus* spp., *Mangifera indica* and *Phoenix dactilifera*.

Introduction of soil, insects, cultures of bacteria and fungi, plant debris, remainder of agricultural products for ship or airplane use, and seed or plant parts for planting purposes if mixed with soil, plants, agricultural products or other materials is prohibited.

Introduction restricted. The following seed can be imported, but require prior approval of the Ministry of Agriculture.

A. Field crops, such as wheat, barley, rice, sorghum, maize, sweet corn, sudan grass, millet, peanut, faba bean, chickpea, onion, sesame, flax, kenaf and castorbean.

B. Vegetable and medical crops, such as pea, bean, cowpea, broadbean, watermelon, melon, cucumber, squash, pumpkin, beet, chard, cabbage, cauliflower, turnip, rocket, radish, leek, lettuce, spinach, endive, chicory, Jew's mallow, hollyhock, tomato, eggplant, pepper, okra, carrot, celery, parsley, dill, anise, common fennel, cumin and coriander.

Not restricted. Importation of other kinds not mentioned above, are not restricted.

Importing potato seed for summer planting

Potato seed imported into Egypt for summer planting must meet the following requirements (Anonymous, 1967):

1. Potato seeds must be of the following varieties: Alplra, Arran Banner, Claudia, Hansa, Giewont, Ker Pondy, King Edward, Patrones, Pier Wiosnek, Eiegline and up-to-date.
2. Import from Holland must be only of Certified classes E, A, B; from Denmark only E or A; from the United Kingdom and Ireland, only class A; from other countries, equal to these classes of seed.
3. Entry of potato seed is prohibited unless they are 100% free from insect infestation of *Popillia japonica*, *Liptinotarsa decemlineata*, *Phthorimea merculella*, and free from disease causing organisms such as *Synchytrium endobioticum*, *Heterodera rostochiensis*, *Ditylenchus destructor*, *Meloidogyne* spp., *Clavibacter michiganensis* subsp. *sepedonicus*, *Pseudomonas solanacearum*, *Erwinia carotovora*, *Erwinia carotovora* subsp. *atroseptica*, leafroll virus, corky ring spot virus, caky ring spot and yellow-dwarf virus.

4. Entry of potato seed may be permitted if the infection of the following pathogens does not exceed 0.5%: *Phytophthora infestans*, *Alternaria solani*, *Fusarium* spp., and *Sclerotium rolfsii.* Internal brown or black spotting, and mechanical damage should also not be more the 0.5%.
5. Potato seed must meet international requirements for importation, as to viruses X, Y, A, M, and leaf curl.
6. Potato seed infected with *Streptomyces scabies* or *Spongospora subterranea* may be imported if they have proper certification documents from the originating country.
7. Potato seed should be free from moderate to severe infection of *Rhizoctonia solani*, but they may be imported with only slight infection if the infected potatoes do not exceed 20% of the lot.

General legislation for export

Egypt's regulations for export of seed, seedlings, fruits, etc., are as follows:

1. Cotton seed may not be exported, whether for planting or industrial use. If any trace of cotton seed is found combined with other seed or plant parts for export, the consignment may not be exported.
2. Export of seed, vegetative seedlings, and other plant parts for planting or breeding purposes must be approved by the Administration of Marketing of Agricultural Crops, Ministry of Agriculture.
3. Export of peanut seed requires a phytosanitary certificate to assure that they are free of aflatoxins produced by *Aspergillus flavus.* Seed must be tested for toxic substances at the Mycotoxin Laboratory. Twenty ppm is the permitted level for human consumption.
4. Potatoes for export to Western Europe and England must have a certificate that they are free from the bacterial brown rot pathogen, *Pseudomonas solanacearum.*
5. Bean seed for export require a phytosanitary certificate that it is free from diseases and insects, particularly bacterial blight disease, *Curtobacterium flaccumfaciens* pv. *flaccumfaciens.*
6. Export to Italy of fruits of the Solanaceae is not permitted except during the period January 1 to February 28, and the fruits must be free of cotton worm and bacterial and viral diseases.
7. Export of potato tubers is prohibited to England, Sweden, Denmark, Austria, West Germany and other countries having the same legislation, unless they are free of potato tuber larvae, soft rot and blight diseases. To facilitate export, potatoes infected with *Alternaria solani* and *Phytophthora infestans* are allowed if infection does not exceed 0.2%. In countries where legislation does not prevent importing these pests or diseases, potatoes may be exported if infection does not exceed 1%.

Provisional seed movement and quarantine regulations

Provisional regulations are those enacted in addition to the basic quarantine regulations.

Specific measures for seed import

1. An import permit is required for seed of restricted category.
2. A phytosanitary certificate issued by the exporting country is required at the port of entry. Phytosanitary certificates should show that seed of bean, lima bean, cowpea, soybean, squash, lettuce, muskmelon are free from viruses.
   Seed imported for breeding purposes may not require a phytosanitary certificate if it does not exceed 0.5 kg in weight upon entry.
3. Imported seed is subjected to direct inspection for contaminants and insects upon arrival.
4. Representative samples are drawn from seed lots and sent to research institutes for seed health testing.

   A. The seed consignment may then be released if it is free of prohibited disease-causing agents and insects.

   B. The seed consignment may require heat treatment (hot water, vapour) or treatment with specific chemicals and fumigants when infected with exotic pathogens and pests.

Specific measures for seed export

The Egyptian government has special stations for screening, handling and controlling many harmful pests and diseases for export purposes. Plants and planting materials which do not meet relevant quarantine legislation cannot be exported.

1. Prior to export, seed must be treated by chemicals recommended by the importing country.
2. Inspection of seed must be done within 7 days prior to export. If the seeds are found infested with viable insects, even in trace amounts, the seed consignment must be fumigated and reinspected after fumigation.
3. If seed is found to be free from insects and disease infection as required by legislation of the importing country, export is permitted, with the following exceptions:

   A. Potatoes for export to England is permitted if the infection of late and early blight pathogens does not exceed 0.2%.

   B. Seed for planting or breeding, if less than 20 kg weight or carried by passengers, does not have to undergo quarantine examination.

General quarantine legislation in the countries of the Near East

Egyptian quarantine documents of 1988 show the following legislation in the countries of the Near East.

Cyprus

Infection of *Streptomyces* spp. and *Spongospora* spp. on potato tuber surface should not exceed 10% of potato tuber surface. Infection should not exceed 0.5% of *Chrysomphallus ficus* on *Citrus* spp., *Phthoremiea operculella*, *Alternaria solani* and *Phytophthora infestans* on *Lycopersicon*, and *Lepidasaphes beckii* on *Citrus*. Peanut seed must be fumigated in the country of export even if not infested. The gas used and application method must be indicated on the phytosanitary certificate. Banana must undergo the same treatment.

Iraq

Import of any planting material or products of date palm (*Phoenix dactaliferea*) is prohibited.

Pakistan

Phoma lingam on *Brassica oleracea* has been prohibited.

Saudi Arabia

Import of planting material or plant parts of date palm including ornamental palms, are prohibited from regions where diseases caused by *Fusarium oxysporum* var. *albedins* are reported. Import is recommended from disease-free countries, or countries provisionally having prior permit from the Saudi Government's Administration of Agricultural Research, Ministry of Agriculture and Water. This date palm disease is common in Mauritius, Algeria, and Morocco. Recently, Morocco has destroyed tens of millions of infected date palms.

All seed for agricultural purposes except wheat seed are warranted for import, but an import permit is required.

In all cases, plant materials cannot be held in, or contain, soil and sand; peatmoss is recommended.

Syria

Prohibited pests are *Pseudococcus citri* on Citrus; *Ceratitis capitata*, *Heterodera rostochinensis*, and *Fusarium oxysporum* on *Lycopersicon esculentum*, *Agrobacterium tumefaciens* or crown gall pathogens on stone fruits, *Synchytrium endobioticum* or blackwart on potato, *Urocystis cepulae* on *Allium sepae*, and *Rhizoctonia solani*.

Turkey

Import is prohibited on plant materials infected with *Chrysomphalus ficus*, *Lepidosaphes beckii*, *Phoma citricarpa*, *Xanthomonas campetris* pv. *citri*, *Agrobacterium tumefaciens*, *Erwinia amylovora*, *Erwinia carotovora*, *Gnorimoschema operculella*, *Phytophthora infestans*, *Pseudomonas syringae* pv. *syringae*, *Pseudomonas syringae* pv. *morsprunorum*, *Phytophthora cambivora*, *Clavibacter michiganensis* subsp. *sepedonicus*, *Synchytrium endobioticum*, *Spongospora subterranea*, *Leptinotarsa decemelineata*, *Heterodera* spp., , *Epetrix cucumeris*, and *Popillia japonica*.

Turkey has issued specific prohibitions on imports infected with the following specified pathogens, although they may not have been named in the Plant Quarantine Regulations (Neergaard, 1980): *Curtobacterium flaccumfaciens* pv. *flaccumfaciens* on *Phaseolus*; *Clavibacter michiganensis* subsp. *insidiosus* on *Medicago sativa*; *Clavibacter michiganensis* subsp. *michiganensis* on *Lycopersicon esculentum*; *Erwinia stewartii* on *Zea mays*; *Pseudomonas syringae* pv. *pisi* on *Pisum*; *Xanthomonas campestris* pv. *campestris* on *Brassica oleracea*; *Xanthomonas campestris* pv. *oryzae* on *Oryza sativa*; *Xanthomonas campestris* pv. *phaseoli* on *Phaseolus*; *Xanthomonas campestris* pv. *vesicatoria* on *Capsicum* or *Lycopersicon esculentum*; *Alternaria brassicae* on *Brassica oleracea*; *Alternaria brassicicola* on *Brassica oleracea*; *Peronospora tabacina* on *Nicotiana*; *Phoma lingam* on *Brassica oleracea*; *Tilletia controversa* on *Agropyron* spp., *Secale* or *Triticum*; *Tilletia secales* on *Secale*; Barley stripe mosaic virus on *Hordeum*, *Secale* or *Triticum*; Soybean mosaic virus on *Glycine max*; Sunflower mosaic virus on *Helianthus annus*; *Anguina tritici* on *Triticum*; *Ditylenchus destructor* on *Allium*.

Categorization of quarantine pests and pathogens

Plant pests and pathogens, whether or not already occurring in Egypt, are subject to quarantine measures. These pests and pathogens have been divided into three categories:

A. Exotic pests and pathogens not present in Egypt: These pests and pathogens are of great epidemic potential, so import of infected or infested host plants and plant products is completely prohibited. These include 62 pests, 79 disease-causing fungi, bacteria and virus and 10 nematodes. Table 1 lists such pathogens, of which most are seed-borne.

B. Pests and pathogens present in Egypt, but prohibited from entry: Seventeen plant pests and 20 pathogens are completely prohibited from entry. Pathogens belonging to this category are listed in Table 2, and most of them are seed-borne.

C. Pests present in Egypt, permitted for introduction provisionally: this includes 90 pests. Plant materials should undergo treatment and/or fumigation.

Table 1. Exotic pathogens not present in Egypt, introduction completely prohibited

Pathogen	Disease and/or Host
Pseudomonas syringae pv. *syringae*	Citrus blast & black pit of lemon
Xanthomonas campestris pv. *citri*	Citrus canker
Erwinia amylovora	Fire bight of pear and apple
Phyllosticta solitaria	Apple
Glomerella cingulata	Apple
Pseudomonas syringae pv. *syringae*	Canker and die back of cherry
Xanthomonas campestris pv. *pruni*	Bacterial spot of plum and peach
Glaeodes pomigena	Plum
Agrobacterium tumefaciens	Crown gall of plum
Pseudomonas syringae pv. *morsprunorum*	Dieback or shot hob of plum
Taphrina pruni	Plum
Dibotryon morbosum	Plum
Pseudomonas solanacearum	Moke disease (bacterial wilt of banana
Pseudomonas syringae pv. *savastanoi*	Knot of olive
Gloesporium olivaum	Anthracnose of olive
Elsinoe ampelinashea	Anthracnose of vine (grape)
Guignardia bidwellii	Black rot of vine
Cryptosporella viticola	Dead arm of vine
Colletotrichum gloeosparioides	Anthracnose of mango
Elsinoe mangifera	Scab of mango
Cephalcuras vorescens	Red rust of mango
Sclerotinia fructigena	Brown rot of stone fruits
Sclerotinia lapa	Brown rot of stone fruits
Sclerotinia fructicola	Brown rot of stone fruits
Aspergillus niger	Smut of fig
Phytophthora infestans	Late blight of fig
Nectria cinnabarina	Dieback of fig
Phomopsis cinerescens	Canker of fig
Yellow dwarf virus	Potato
Corky ring spot virus	Potato
Synchytrium endobioticum	Blackwart of potato
Clavibacter michiganesis subsp. *sepedonicus*	Ring rot of potato
Spindle tuber virus	Potato
Xanthomonas albilineans	Gumming disease of sugarcane
Xanthomonas campestris pv. *vasculorum*	Gumming disease of sugarcane
Phytophthora erythrosepteca	Phytophthora rot of sugarcane
Puccinia kucherii	Rust of sugarcane
Helminthosporium sacchari	Helminthosporium eye spot of sugarcane
Helminthosporium stenospiliem	Sugarcane
Diplodia zeae	Diplodia stalk and ear rot of maize
Gibberella zeae	Gibberella rot of maize
Erwinia stewartii	Stewart's disease (bacterial wilt) of maize
Helminthosporium maydis	Leaf spot of maize
Helminthosporium carbonum	Leaf spot of maize
Cochliobolus heterostrophus	Leaf spot of maize
Sphacelotheca crienta	Loose kernel smut of sorghum

.....cont.

Pathogen	Disease and/or host
Pythium arrhenomanes	Pythium root rot of sorghum
Claviceps purpurea	Ergot of wheat
Gibberella spp.	Gibberella scab of wheat
Fusarium spp.	Fusarium scab of wheat
Ophiobolus graminis	Take-all of wheat
Helminthosporium sativum	Crown rot and root rot of wheat
Septoria tritici	Septoria leaf blotch & glume blotch of wheat
Septoria nodorum	Septoria leaf blotch & glume blotch of wheat
Typhula itoana	Typhula blight of wheat
Typhula idahoensis	Typhula blight of wheat
Tilletia caries	Bunt (stinking smut) of barley
Septoria passerinii	Septoria leaf blotch of barley
Xanthomonas campestris pv. *translucens*	Black chaff of barley
Thielaviopsis basicola	Black root rot of barley
Colletotrichum destructivum	Anthracnose of tobacco
Cercospora oryzae	Narrow brown leaf spot of rice
Ophiobolus oryzinus	Arkansas footrot of rice
Gibberella fujikuroi	Bakanae disease of rice
Neovossia horida	Black smut of rice
Ustilaginoidea virens	False smut of rice
Colletotrichum wicinans	Smudge of onion
Botrytis gladiolorum	Botrytis rot of gladiolus
Botrytis tulipae	Botrytis of tulip
Botrytis convoluta	Botrytis of iris
Gleossporium thumenii	Anthracnose of tulip
Phytophthora cactorum	Flower blight of tulip
Rhizoctonia tuliparum	Grey mould of tulip
Fusarium bulbigenum	Basal rot
Stagnospora curtisii	Leaf scorch
Ramularia vallisumbrosae	White mould
Botrytis narcissicala	Smoulder of narcissus
Phytophthora castorum	Stem basal rot of lily
Fusarium oxysporum	Fusarium rot of lily
Botrytis hyacinthi	Botrytis rot of hyacinth
Heterodera schactii	Cyst nematode on beetroot & other crops
Heterodera rostochiensis	Cyst nematode on potato
Pratylenchus pratensis	Root lesion nematode on potato
Ditylenchus destructor	Potato rot nematode
Aphelenchoides fragaria	Spring crimp nematode on strawberry
Anguina tritici	Gall nematode on wheat
Ditylenchus dipsaci	Bulb and stem nematode on onion
Ditylenchus angustus	Stem nematode on rice
Hirschmanniella oryzae	Root nematode on rice
Meloidogyne spp.	Root knot nematode on many crops

Table 2. Pathogens present in Egypt, but introduction completely prohibited

Pathogen	Disease and/or Host
Pyricularia oryzae	Blast of rice
Helminthosporium oryzae	Brown spot of rice
Entyloma oryzae	Leaf smut of rice
Urocystis agropyri	Flag smut of wheat
Helminthosporum turcicum	Leaf blight of sorghum
Sphacelotheca reiliana	Head smut of sorghum
Tolyposporium chrenbergii	Long smut of sorghum
Cephalosporium acremonium	Black bundle of maize
Helminthosporium turcicum	Leaf blight of maize
Ustilago maydis	Corn smut of maize
Sphacelotheca reiliana	Head smut of maize
Ustilago scitaminea	Smut of sugarcane
Diplodia phoenicum	Diplodia rot of palm
Venturia inaequalis	Scab of apple
Venturia pyrina	Scab of pear
Taphrina deformans	Scab of peach
Puccinia pruni spinorae	Rust of stone fruits
Erwinia carotovora	Bacterial soft rot of potato
Erwinia carotovora subsp. *atroseptica*	Blackleg of potato
Pseudomonas solanacearum	Brown rot of potato

Seed health testing procedures

The Egyptian Quarantine Service sends a large number of seed samples of different crops to the ARC Seed Pathology Department. The Seed Pathology Department has 4 trained staff in seed health testing. The ARC also has one trained staff each at the Horticultural Crop Disease Department, Viral Disease Department, and Legume Disease Department. The Department conducts seed health testing using internationally-standardized methods.

In general all seed samples are subjected to inspection of dry seed by naked eyes, primarily for detecting prohibited nematode galls and fungal sclerotia. Special attention is paid to cereal seeds in this test. Washing test is also conducted routinely as it gives quick information on presence or absence of bunts or smuts in wheat and barley, the blast fungus in rice, and many *Drechslera* and *Botrytis* spp. in seeds of a number of crops. The common incubation method is the blotter method while agar plating is conducted when required for special types of seed. Indicator plant tests are conducted in greenhouses only for detecting potato viruses. And post-entry quarantine in closed glasshouses is limited to the port of Alexandria for detecting dangerous fungal, bacterial and viral pathogens. Special tests such as the embryo-count method, biochemical test, phage-plaque and serological methods are only performed in very special cases.

A part of the sample is tested at the Division of Nematode Research if nematodes are suspected. Testing for nematodes is usually done by washing the seeds. Similarly, the sections of Bacterial and Viral Plant Diseases may also examine part of the samples, especially when the number in each sample is large.

In most developing countries, availability of testing facilities is a limiting factor. And even if some testing facilities are present, they are insufficient. The other major difficulty encountered is the non-availability of trained manpower in seed health testing techniques. This, in fact, is one of the main reasons why in Egypt quarantine service sends the imported materials to specialized laboratories. The present staff of the quarantine service is not even familiar with the identification of quarantine organisms.

Another difficulty, which is encountered, is the testing of pesticide treated seeds for seed health.

Conclusion

The list of seed-borne diseases of quarantine significance in Egypt was established long ago. It is time to revise it based on present distribution of dangerous plant diseases in Near East countries.

Quarantine facilities at different stations of Egypt are insufficient both in terms of testing equipment and trained manpower, including post-entry growing-on facilities. The governmental authorities must give serious considerations to such deficiencies. Efforts should be made to survey the existing plant diseases in the country and their distribution in different parts which is the basis of making realistic quarantine regulations. The entry of plants and germplasm must be regulated and new approaches to control diseases, especially against viruses in germplasm must be worked out.

References

Abdelmonem, A.M., El-Wakil, A.A., Shaarawy, M.A. and Mathur, S.B. 1989. Investigations on transmission of rice diseases by seeds and principles for their control. I. Fungi associated with rice seed and some observations on seed-borne infections. Paper presented at the *22nd International Seed Testing Congress*. Edinburgh, Scotland. June 21-30.

Anonymous 1967. Al Wakhaea Al Masria, No. 119.

Karpati, J.F. 1981. History and current status of the International Plant Protection Convention (IPPC). Paper presented at the *Seventh Session of the Near East Plant Protection Commission*. FAO. Rome. AGP/NEPPC/81/WP2.

Karpati, J.F. 1983. Plant quarantine on a global basis. *Seed Science and Technology* **11:** 1145-1157.

Leppik, E.E. 1962. Distribution of downy mildew and some other seed-borne pathogens on sunflowers. *FAO Plant Protection Bulletin* **10:** 126-129.

Leppik, E.E. 1964. Some epiphytotic aspects of squash mosaic. *Plant Disease Reporter* **84:** 41-42.

Locke, C.M. and Watson, A.I. 1955. Foreign plant diseases intercepted in quarantine inspection. Plant Disease Reporter **38:** 518.

Neergaard, P. 1979. *Seed Pathology*. Vol. I & II. The Macmillan Press Ltd. London and Basingstoke. 1191 pp.

Neergaard, P. 1980. A review on quarantine for seed. Nat. Academy of Science, India. Golden Jubilee Commemoration Volume: 495-530.

Neergaard, P. 1984. Seed health in relation to exchange of germplasm: Seed management techniques for genebanks. *Proc. Workshop Royal Botanic Gardens*. Kew: 1-18.

Reddy, D.B. 1970. Loose smut on wheat (in loads). *FAO Plant Protection Bulletin* **18:** 147.

Quarantine for Seed in the United States *

WALTER J. KAISER

USDA, ARS, Regional Plant Introduction Station, Washington State University, Pullman, Washington 99164-6402, USA

Introduction

In the last twenty years, there has been a large increase in the flow of plant materials between and within countries. A significant portion of this material is vital to the crop improvement programmes of many countries. Large germplasm collections, often referred to as gene banks, have been established for different food crops in many countries in recent years (Plucknett *et al.*, 1983).

Most food, forage, and fiber crops are propagated by true seeds (Neergaard, 1977). Many plant pathogens, including bacteria, fungi, viruses, viroids, and nematodes affecting these seed-propagated crops are transmitted in, on, or with seeds. Seed transmission is an efficient method of distributing plant pathogens through time (carry over from season to season) and space (spread from one location to another). Seed transmission has been responsible for the introduction of a number of plant pathogens into different countries (Agarwal and Sinclair, 1987; Baker, 1972; Neergaard, 1977).

Plant introduction in the United States

Plant introduction has played a very important role in the development of a strong and diversified agriculture in the United States. Most of the major food, fiber, and forage crops presently cultivated in the United States originated elsewhere and were introduced over the years by various means (Waterworth and White, 1982). Presently, most formal plant introduction activities in the country are associated with the National Plant Germplasm System (NPGS) which is a coordinated network of federal, state, and private institutions in the United States working cooperatively to acquire, increase, evaluate, document, maintain, enhance, and distribute plant genetic resources. The four Regional Plant Introduction Stations (RPIS) located at Ames, Iowa; Geneva, New York; Griffin, Georgia; and Pullman, Washington are important

* Paper not presented at the Workshop

components of the NPGS. The Regional Stations maintain germplasm of true seeds of more than 140,000 plant accessions in over 900 genera and 4,600 species (White *et al.*, 1989). Germplasm collections of a few important food crops, like wheat (*Triticum aestivum*), peanut (*Arachis* spp.), potato (*Solanum* spp.), soybean (*Glycine max*), and cotton (*Gossypium* spp.) are not located at the RPIS, but are maintained by crop-specific curators. For example, the National Small Grains Collection, with over 110,000 accessions of wheat, barley, oats, rice, rye, triticale, and *Aegilops*, is located at Aberdeen, Idaho. The National Seed Storage Laboratory at Fort Collins, Colorado is responsible for long term storage and preservation of seeds of a broad range of domestic and foreign plant introductions.

About 9,000 new accessions are added to the NPGS annually (White *et al.*, 1989). The transfer of such large amounts of germplasm, however, is not without its dangers, particularly when it results in the introduction of exotic pathogens or new strains or races of indigenous pathogens (Kaiser, 1983). Not only can new pathogens be introduced, but germplasm collections may serve as reservoirs of different seed-borne plant pathogens.

Plant quarantine regulations

In the United States, four federal laws authorize appropriate actions to prevent the introduction and spread of plant pests into and within the country and measures to control or eradicate indigenous or foreign pests (Parliman and White, 1985; White *et al.*, 1989).

a) Plant Quarantine Act of 1912 is the first federal law enacted in the United States that deals with plant quarantine matters. It authorizes federal involvement in activities to control the introduction of exotic pests and to implement domestic quarantines to prevent the spread of pests new to or not widely distributed in the United States.

b) Organic Act of 1944 authorizes pest surveys and control programmes against both indigenous and exotic pests. The act authorizes the issuance of phytosanitary certificates in accordance with the requirements of importing states and foreign countries.

c) Federal Plant Pest Act of 1957 authorizes emergency actions to prevent or delay the introduction and interstate movement of plant pests not covered by the Plant Quarantine Act of 1912. The Act of 1957 broadly defines plant pests as, "any insects, mites, nematodes, slugs, snails, protozoa, or other invertebrate animals, bacteria, fungi, other parasitic plants or reproductive parts thereof, viruses, or any organisms similar to or allied with any of the foregoing or any infectious substances, which can directly or indirectly injure or cause disease or damage in any plants or parts thereof, or any processed, manufactured, or other products of plants" (Rohwer, 1979). The act also authorizes importation of plant pests into and within the United States for scientific study.

d) The Federal Noxious Weed Act of 1974 authorizes the introduction of known noxious weeds from other countries for research and germplasm programmes. Certain noxious weeds are important components of some germplasm collections and are used for research purposes, particularly in breeding programmes. Scientists must obtain permits to obtain seeds of noxious weeds from U.S. germplasm collections or from other countries.

Passage of these four acts has resulted in the issuance of numerous plant quarantine regulations which are contained in the Code of Federal Regulations under Title 7 (Agriculture), Chapter III (Animal and Plant Health Inspection Service (APHIS, USDA). Many of the regulations pertaining to the importation of seeds or vegetative propagules of most plant species are found in parts 319.37-1 to 391.37-14 (Nursery Stock, Plants, Roots, Bulbs, Seeds, and Other Plant Products) of the Code (USDA, 1989). These regulations are amended and revised regularly. The federal plant quarantine regulations are administered by the Plant Protection and Quarantine Programme of APHIS.

Quarantine categories for imported seeds

Germplasm of most plant species imported into the United States for research purposes falls into one of three quarantine categories: restricted, postentry, and prohibited (Foster, 1981; USDA, 1989; Waterworth, 1981; Waterworth and White, 1982; White *et al.*, 1989). These categories are based on the disease and pest situation in the country of origin and the importation of these different plant species which would have potential danger to agriculture in the United States.

The quarantine regulations governing the entry of vegetatively propagated plants (e.g., potatoes, fruit trees) are usually more stringent that for those propagated by true seeds. Most germplasm imported as seeds is in the restricted (least restrictive) category. Many vegetatively propagated crop species fall into the postentry and prohibited categories which require importation permits, predesignated growing sites or special testing procedures in quarantine to detect latent infections by potentially important exotic pathogens, particularly viruses and viroids.

Most forage, vegetable, and flower species imported as seed into the United States are in the restricted category. Importation of plant species in this category generally does not pose a threat to the country's agriculture.

An import permit may be required for seeds of some genera in the restricted category before they can be imported legally into the United States. Applications for a written import permit should be submitted to the Permit Unit, PPQ, APHIS, USDA, Federal Building, Hyattsville, Maryland 20782 at least 30 days prior to arrival of the article at the port of entry. Application forms can be obtained from the Permit Unit, PPQ, APHIS in Hyattsville, Maryland or from local PPQ, APHIS offices. Seed may or may not be subjected to visual inspection or microscopic examination at one of the inspection stations at specified ports of entry located near the U.S. border.

Depending on the pest situation in the country of origin, seeds of some plant species may require treatment with specific chemicals effective in controlling certain pests (Table 1). Once these requirements are met, there are few, if any, regulations on how or where the seeds can be grown.

Table 1. Examples of crops imported as seeds from certain countries or regions to the United States that must be treated with pesticides before release.

Genus	Country of origin	Treatment
Lathyrus spp. (sweet pea) *Lens* spp. (lentil) *Vicia* spp. (faba bean, vetch)	All countries, except those in North and Central America	Fumigation to control insects of the family Bruchidae
Glycine spp. (soybean) *Dolichos* spp. (lablab) *Pachyrhizus* spp. (yam bean) *Phaseolus* spp. (bean)	Africa, Australia, Brazil, Burma, Cambodia, China, Costa Rica, India, Indonesia, Japan, Korea, Laos, Malaysia, Nepal, New Caledonia, Papua New Guinea, Philippines, Sri Lanka, Taiwan, Thailand, USSR, Venezuela, Vietnam, West Indies	Treat seeds with Patterson's Multi-purpose fungicide (a zineb-captan formulation) for control of infection by *Phakopsora pachyrhizi* (soybean rust) "
Vigna spp. (cowpea, adzuki bean)	"	"
Medicago spp. (alfalfa)	Europe	Dust seeds with thiram (50%) for control of infection by *Verticillium albo-atrum*

Seed of a few plant genera originating from certain countries or regions of the world are placed in the prohibited quarantine category. Importation of seeds in the prohibited category poses a potentially serious threat to U.S. agriculture. Certain seed-borne pests (or their strains or races) in the exporting countries may not be present or widely distributed in the United States and, if introduced, would have the potential for spreading rapidly and causing serious crop losses. The imported plant materials also may serve as alternate hosts of indigenous pests or they may be infected with more virulent races or strains of particular pests not presently established or widely distributed in the United States. Some of the genera included in the prohibited category are lentil (*Lens* spp.), maize (*Zea* spp.) rice (*Oryza* spp.), cotton (*Gossypium* spp.), potato (*Solanum* spp.), and wheat (*Triticum* spp.) (USDA, 1989) (Table 2). Seeds in the prohibited category may, nevertheless, be imported with a special Departmental permit for scientific purposes provided safeguards are adequate.

Regulations may vary depending on the country of origin of the seed. For example, lentil seed from South America is on the prohibited list because of *Uromyces viciae-fabae* which occurs in this region of the Americas, but not in the United States (USDA, 1989). Seeds of lentil germplasm from South America that are being imported for research purposes with a special Departmental permit are inspected at the Plant Germplasm Quarantine Center, Beltsville Agricultural Research Center East, Beltsville, Maryland 20705 before release to the scientist.

Table 2. Some major crops (seeds) from certain countries or regions of the world that are in the prohibited quarantine category of the United States.

Crop	Country of origin	Disease
Lens spp. (lentil)	South America	Rust
Oryza spp. (rice)	All countries	Several bacterial, fungal and viral diseases
Zea spp. (maize)	Asia, Africa, Australia, New Zealand, Oceania	Downy mildews, *Physoderma* spp.
Triticum spp. (Wheat)	Afghanistan, India, Iraq, Mexico, and Pakistan	Karnal bunt
	Afghanistan, Algeria, Australia, Bangladesh, Bulgaria, Chile, China, Cyprus, Egypt, Falkland Islands, Greece, Guatemala, Hungary, India, Iran, Iraq, Israel, Italy, Japan, Korea, Libya, Morocco, Nepal, Oman, Pakistan, Portugal, Romania, Spain, Tanzania, Tunisia, Turkey, South Africa, USSR, and Venezuela	Flag smut

State quarantines

Several states in the United States also have implemented plant quarantine regulations. A few states had promulgated plant quarantine regulations before the first federal plant quarantine act was passed in 1912. Most states enact plant quarantine laws to meet their own needs, particularly where there is no domestic federal quarantine or there is a need to solve a local problem (Rosenberg, 1989). However, federal domestic quarantine regulations preempt any state quarantine regulations for the same pest problem. At times, federal and state quarantine programmes will join forces to control a specific pest problem.

California was the first state to enact quarantine regulations. In 1885, the state enacted quarantine legislation to prevent the introduction of grape phylloxera (*Phylloxera viticola*) and other pests on imported grapevines.

Seed pathology research in the NPGS

Research on seed-borne diseases of plant germplasm in the NPGS is a responsibility of the plant pathologists associated with the four Regional Plant Introduction Stations. Kaiser (1987) has summarized the type of seed pathology research that is conducted at the different regional stations. Pioneering work on seed-borne diseases of plant germplasm was initiated by Leppik in the late 1950s and 1960s while he was at the Ames RPIS (Kaiser, 1987). Studies by Leppik and other RPIS plant pathologists in the last 30 years have detected and identified several potentially important pathogens on seed of imported germplasm, particularly the large-seeded food legumes like pea, bean, and lentil. Improved methods of detecting seed-borne pathogens, such as enzyme-linked immunosorbent assay (ELISA), has aided in the detection of seed-borne viruses in imported germplasm. This new technology is being used at the Pullman RPIS to eradicate seed-borne bean common mosaic virus from the *Phaseolus* (bean) germplasm collection and to free the *Pisum* (pea) germplasm collection at the Geneva RPIS of pea seed-borne mosaic virus (Kaiser, 1987).

References

Agarwal, V.K. and Sinclair, J.B. 1987. *Principles of Seed Pathology*. Vol. 2. CRC Press, Boca Raton, Florida.

Baker, K.F. 1972. Seed pathology. pp. 317-416. In: *Seed Biology*. Vol. 2. (ed. T. Kozlowski). Academic Press, New York.

Foster, J.A. 1981. Regulation of plant germplasm imported into the United States. *EPPO Bulletin* **11:** 155-162.

Kaiser, W.J. 1983. Plant introduction and related seed pathology research in the United States. *Seed Science and Technology* **11:** 1197-1212.

Kaiser, W.J. 1987. *Testing and Production of Healthy Plant Germplasm*. Technical Bulletin No. **2**. Danish Government Institute of Seed Pathology for Developing Countries, Copenhagen, Denmark. 30 pp.

Neergaard, P. 1977. Seed pathology. Vol. 1 & 2. The Macmillan Press, London and Basingstoke.

Parliman, B.J. and White, G.A. 1985. The plant introduction and quarantine system of the United States. pp. 361-434. In: *Plant Breeding Reviews*. Vol. III. (ed. J. Janick). Avi Publishing Company, Wesport, Connecticut.

Plucknett, D.L., Smith, N.J.H., Williams, J.T., and Murthi Anishetty, N. 1983. Crop germplasm conservation and developing countries. *Science* **220:** 163-169.

Rohwer, G.G. 1979. Plant quarantine philosophy in the United States. pp. 23-34. In: *Plant Health, the Scientific Basis for Administrative Control of Plant Diseases and Pests*. (eds. D.L. Ebbels and J.E. King). Blackwell Scientific Publications, Oxford, England.

Rosenberg, D.Y. 1989. The interaction of state and federal quarantines in the U.S. pp. 59-74. In: *Plant Protection and Quarantine*. Vol. III. (ed. R.P. Kahn). CRC Press, Boca Raton, Florida.

USDA (United States Department of Agriculture). 1989. Code of Federal Regulations, Title 7, Agriculture, Chapter II, Animal and Health Inspection Service, Parts 300 to 399, Washington D.C. U.S.A.

Waterworth, H.E. 1981. Control of plant diseases by exclusion: Quarantines and disease-free stock. p 269-296. In: *Handbook of Pest Management in Agriculture*. Vol. 1. (ed. D. Pimentel). CRC Press, Boca Raton, Florida.

Watherworth, H.E. and White, G.A. 1982. Plant introductions and quarantine: The need for both. *Plant Disease* **66:** 87-90.

White, G.A., Shands, H.L. and Lovell, G.R. 1989. History and operation of the National Plant Germplasm System. pp 5-56. In: *Plant Breeding Reviews*. Vol. 7. (ed. J. Janick). Timber Press, Portland, Oregon.

Precautions Directed against Introduction of Pests with Seeds into Uninfected Areas

R. IKIN

Plant Pathology and Quarantine Group, Plant Production and Protection Division, FAO, Rome, Italy.

Introduction

There is a need for recognizing the importance of effective plant quarantine procedures because increasingly new pests and diseases (collectively referred to in this paper as pests) are being recorded for the first time in new locations. The only feasible way they could have moved to new areas is with seed.

This paper explains the sequence of events that should take place when one takes precautions against the spread of pests associated with seed. The types of seed transmission, and examples of particular pests are given elsewhere in this paper and will only deal with the philosophies behind the operation of plant quarantine procedures.

Definitions

Firstly, we should recognize that pests can be spread in various ways. There are differences in terminology that is often confused unwittingly. These terms are seed contamination, seed-borne, and seed transmission. But from a quarantine point of view these differences are important.

Seed contamination

Many pests can be associated with seed, and they can be non-specific, some are storage pests that are normally thought of as problems of trade in plant commodities e.g. khapra beetle, but given the right conditions they can be a problem. Of course, seed that is unclean and mixed with weed seeds and soil should not be permitted entry without treatment to eliminate these most obvious pests. Indeed there is an obligation on the part of a seed collector/curator to ensure that seed lots exported are not likely to harbour any pests, for example not harvesting seed from obviously diseased plants (this is an International Plant Protection Convention obligation when issuing a Phytosanitary Certificate).

Seed-borne

Many pests have been shown to be present on or in the seed at some stage of the life cycle, but because the pest is not capable of being sustained until the vegetative (plant) stage they cannot be perpetuated in this manner. Normally these pests can be eliminated by a treatment with a systemic or protective chemical.

Seed transmission

This category includes those pests that have been found to systemically infect seeds and which if the seed is sown continue the association into the next life cycle stage. Seed transmission rates vary from plant to plant and from pest to pest. It is rarely 100% and therefore mechanisms can be adopted to select out those seeds that do not carry the pest.

Problems of the movement of seeds

Seed has been responsible for transporting a number of important pests worldwide. In fact a number of very important pests of crop plants have been identified as pests which have been perpetuated within seed germplasm collections and which have then been spread from these collection into crops in new locations where they have caused considerable economic losses. Examples of these are potato spindle tuber viroid that was distributed to Europe and Australia through true potato seed collections, pea seed-borne mosaic that was detected in North American pea and bean collections but not before it had also become a field pest in peas and a number of important rice pathogens that have been distributed to various countries with new rice varieties. Even today there is considerable concern at the role of some germplasm banks in the distribution of both pests and the seed, this will be discussed later.

Concepts of risk - Pest Risk Analysis

In considering the procedures to be adopted towards the importation of seed at a national level the most important matter to be resolved is to determine those pests which could be categorized as of "quarantine concern". In its simplest form this could involve the listing of those pests which occur in the country of origin and subtracting this from those that occur in the country of destination. If the countries have the same pest compliments then there should be no impediment to the movement of material. However, this is rarely the case, and in many instances complete lists are not readily available so once a list of exotic pests is identified further consideration has to be given to the potential economic importance that might result from the introduction of one or more of these pests. Introduction in this case means introduction and establishment. Once this activity is completed then the matter of suitable and effective treatments should be considered to determine if the risk assessed as significant by the previous steps in the analysis can be addressed to reduce it to acceptable levels. In reaching a regulatory position to either accept the import with specific conditions to be met, or on the other hand, to not accept the level of risk and prohibit import the task of **Pest Risk Analysis** (PRA) has been undertaken. FAO in co-operation with Regional Plant Protection Organizations is currently attempting to rationalize the mechanisms of PRA so that countries using the procedure will come to similar regulatory positions if the pest risk is also similar (Ikin, 1990).

Risk categories for seed

It should therefore be possible in considering the import of a particular crop as seed, to categorize the risks for the range of pests that a country considers of quarantine concern. In a simplistic form these could be:

Low risk — That could be imported subject only to visual inspection, and treatment only if a quarantine pest is detected.

Medium risk — That could be imported subject to mandatory treatment.

High risk — That could only be imported if grown in post entry quarantine and release only of progeny.

In determining the level of risk there must be considerable inputs from databases on pest distribution. This is sometimes incomplete and assumptions often have to be made. For important commercial crops information is generally available worldwide to a reasonable level of confidence, but with the recent emphasis on the collection of wild and closely related species in areas of the origin of genetic diversity, the lack of information on the pest status of these "Primitive" types combined with the potential for its transfer to their economic crop relatives creates assessment problems.

Before each of these categories are considered in detail some general observations are worthy of mention. In the case of insect pests and to a lesser extent nematodes the availability of fumigation with broad spectrum chemicals enables the pest risk to be effectively eliminated by a single treatment at appropriate concentrations. This is not possible for other pest types. In other cases the importation of seeds from countries, or areas within a country, where the particular pests of concern are not present (pest avoidance and pest free areas) is possible, but not often utilized effectively. Pathogen testing schemes are available in some countries for important commercial crops e.g., bean seed in California, Idaho and Washington State which produce commercial quantities of seed free from a wide range of seed-borne pathogens. The crops are examined visually several times during the season and specific and drastic action taken if a pathogen is detected in a crop destined for export. Such schemes provide crops free of specific diseases and some of them rely on indexing tests to supplement visual screening e.g., ELISA tests for pea seed-borne mosaic virus on *Pisum* in the USA, and avocado sun-blotch viroid in California.

Low risk category. Seed import of this category is subjected to inspection and treatment if necessary. Seed health testing which is carried out by laboratories concerned with seed quality is not necessarily synonymous with the tests that are needed to satisfy plant quarantine requirements, although many of the methods developed by the ISTA are utilized by the plant quarantine discipline. The differences are because the purpose of seed health testing is to assure a particular quality of product that is generally used for local sale and local production, and more often it involves bulk quantities of seed. Certain levels of contamination or infection are acceptable in this case. In the case of plant quarantine the PRA has already determined that those pests that may be detected in the seed are of low risk and may be permitted entry, and that the tests are conducted to ensure conformity with an already established position. If the material is not found to conform then the seeds are reconsigned, destroyed or if acceptable will be treated before release.

Therefore the range of pests tested for in this category cannot be extensive, as determined already by the PRA, and so the laboratory and equipment needs are not as elaborate as those for seed health testing.

Medium risk category. Seed import of this category is subjected to mandatory treatment. The earliest forms of treatment of seed were conducted by farmers who wished for their seedbeds to be clear of infection, and many of these treatments which were found to be effective against both diseases and pests were adopted by plant quarantine services worldwide. Unfortunately, many of them dealt with dusting with mercury or other heavy metal compounds and now are not available, or cannot be recommended. Additionally treatments whose efficacy is acceptable for field control may not be acceptable as quarantine treatments and therefore specific treatments for quarantine purposes have been developed.

A number of treatments for seed are included in the International Plant Quarantine Treatment Manual (IPQTM)(FAO, 1983).

In considering insect pests of seed the most widely used treatment which is recognized as very effective is fumigation, particularly with methyl bromide at normal atmospheric pressure. The gaseous phase of MeBr is very penetrating and given an effective fumigation treatment almost all insects (all insects to 95% confidence limits) will be killed, and at all stages of their life-cycle. Fumigation at reduced pressure (partial vacuum) has been acceptable as treatments for some pests, but the treatment only reduces treatment time and given the complexity of operating vacuum fumigation equipment, its initial cost and the technical inputs needed to maintain the facility this method cannot be recommended for widespread adoption, unless it is necessary to treat internal seed pests.

Fumigation with other chemical is recognized for a number of seeds, particularly when they have been found sensitive to MeBr. These include carbon disulphide and hydrogen cyanide.

In some cases the fumigation is also effective against nematodes, though treatment with hot water is generally recommended for these pests. Treatments that are effective against nematodes are listed in the IPQTM.

High risk category. Seed import of this category must be subjected to post-entry quarantine by growing-on test and only the seeds produced by healthy plants are released.

Despite the availability of many effective seed treatments there is sometimes no effective treatment against a quarantine pest, or the treatments are not as efficient as the pest risk involved. Moreover no seed treatments are available against viruses. In such cases the only recourse then is to grow the plants in post-entry quarantine. The purpose of this procedure is to screen the plants produced from the seed, to reject those found to be infected and to produce seeds from the remainder which can be released as a nuclear stock. In the case of plants which are normally propagated vegetatively they are screened for disease and budwood or cuttings are released from the initial plants which are then destroyed.

Evaluation of the need for post-entry quarantine (PEQ)

As a general principle post-entry quarantine is used only for valuable material of high risk and the quantity of seed of any particular line should be restricted to an amount that is sufficient to establish the line in the importing country. Pest avoidance is a concept that is not widely used when addressing the risk of the introduction of serious pests. It is often possible even within a region known to be infested with a pest to find areas where the pest is not present and the opportunity should be taken to take advantage of this. Imports of large quantities of seed of a single line or of the same genetic compliment defeats the objective of PEQ which relies on the efficiency and accuracy of screening procedures carried out on only small numbers of plants.

The role of PEQ in addressing risk

The task of PEQ is to grow the imported material in some isolation from other susceptible material with the purpose of being able to detect quarantine pests if present, to remove that material which is infected and to produce seeds from the remainder which can be released as disease free (Sheffield, 1968). To undertake this task specific facilities are required, but there is considerable differences of opinion as to the technical and structural complexity of these requirements which in many cases ignore the operational efficiency of already established facilities. The retention of plants in quarantine is in many cases evaluated as a need to maintain plants in a completely enclosed system that will not permit the ingress or egress of any organism, pathogenic or otherwise (Kahn, 1983). This ignores the experience of many countries of successfully executing post-entry quarantine facilities where this type of structure is not available or where its adoption because of climatic conditions is clearly impractical (Smee, 1975; Smee and Setchell, 1973).

PEQ facilities

Except in temperate climates, where winter heating is required I consider that completely enclosed greenhouse facilities are impractical because of the mechanical complexity of the systems in adjusting environment, especially temperature at which plant growth can be sustained. In developing countries the maintenance needs of these facilities is a considerable burden on technical inputs, and the infrastructure costs such as electricity and staffing are prohibitive, if they are ever evaluated.

The need for completely enclosed systems appears to have developed from the adaption of structures used for the containment of biological control agents in the receiving country when on-going evaluation of host specificity is required. The risk associated with this type of operation can be considered similar to PEQ, but the risk factors are not of the same magnitude.

In the case of biological control agents there is continued association between the host and pest over a prolonged period, often at high population pressures in confined spaces where the pest may wish to seek other hosts. The pest is also maintained through all its stages, including that for optimal dispersal.

In the case of PEQ the pest is normally in very low concentration. Once detected it should be removed with its host and destroyed. It is generally present without its vector if one is needed, and when detected is normally in its multiplication and adaption phase not its dispersal phase.

There are some major problems with the operation of enclosed systems. If these completely enclosed greenhouse systems with airconditioning do malfunction, then the environmental change can be so rapid and irreversible, that within a few hours the temperatures rise to levels where the plants in the greenhouse desiccate and are killed. These enclosed systems do not allow for the structures to be ventilated other than by powered fans.

Highly sophisticated air-conditioned greenhouses constructed as PEQ facilities in Nigeria, Zambia and Kenya (Kahn, 1983) are now in such disrepair that they are no longer functional, and will require considerable inputs to restore them to their original specification. Similarly, more recent structures in the Dominican Republic and Jamaica, although able to operate with the necessary inputs, do not do so because the technical personnel needed to operate this complex cannot be found or are not being employed. They remain uncommissioned or operating at minimal levels.

Passive systems which are based on simple design of structures for cooling and ventilation are more acceptable to a developing country. It is true that sometimes all-year-round operation is not possible with such structures, but a controlled environment cabinet can be used whenever required.

Concerning the spread of plant pathogens within green and screenhouses the experience in the maintenance of exotic pathogens in isolation has indicated that a small bench separated from others by distances of half a meter is all that is required.

Universities and research organizations often need to work continuously with exotic pathogens, including viruses, maintained on a long term basis even within a single cubicle. Cross infection, even in these circumstances where daily manipulation of diseases is required, is rare. There is no record of involuntary escape of pathogens, particularly viruses, from infected plants in greenhouses/screenhouses onto crops established in adjacent fields. The essential point is that staff should be able to operate safely if they take the necessary precautions. A number of PEQ stations have been designed and located away from production areas because of this risk. Such isolation has caused difficulties in major administrative and technical support affecting the operation of the stations. It is far more important to have readily available diagnostic capability to quickly identify a suspect pest or pathogen than to have the seemingly important isolation from other crops.

Certainly in the case of facilities which enclose the plants under PEQ conditions further isolation from areas with other related crops seems to be unjustified. It should be a matter of policy to locate PEQ facilities in developing countries close to the location of pathologists, entomologists and horticulturists. Some facilities located in isolation have not functioned effectively because of difficulties in obtaining staff to work in isolation, to obtain first hand attention from distantly located specialists. Such conditions do create excessive risks of spreading pests when suspected material on plant tissue has to be transported to another location for identification. It is, therefore, far better to locate PEQ stations within already existing research facilities as long as the PEQ staff are to some extent independent of research staff. In one case a facility has not been used since it was commissioned in 1977 because of the difficulties in the recruitment of staff to work at the isolated location.

Nevertheless any structure must have certain safety features and these include:

- Insect screening to keep out insects from the local population, and keep those in the structure those that might have been introduced,
- Adequate and enclosed drainage that will retain the waste water within the quarantine area,
- Solid floor to permit regular cleaning of the area,
- Double doors against the ingress of flying insects,
- Facilities for the production of growing media and its disposal, and
- Benches to support pots above ground that can be easily cleaned (galvanized).

Most of these items are covered by Kahn (1983) and Smee and Setchell (1973).

PEQ operations - greenhouses and screenhouses

The purpose of PEQ is to grow the imported seed under acceptable climatic conditions. PEQ should only be used in cases where the assessment of risk (see previous section) is such that other methodologies for the production of healthy seed or plants is not possible. It is a complex and difficult task which needs considerable technical and other inputs and the following general conditions should be observed:

- Small samples of fumigated and treated seed should be sown, only sufficient to ensure the retention of the genetic compliment required of the introduction. Bulk imports of seed for field planting is not really PEQ.
- Seed should be treated before planting with an acceptable dressing in order to ensure that the level of disease is reduced further. Harvested seed should also be treated before release, registered chemicals only should be used.

PEQ is done in order that:

- Diseases that may be present are evident so that infected plants can be removed and destroyed. The following points should be considered:
    - Conditions for the expression of the pest may need manipulation of the environment; high moisture and injury promote the expression of bacterial diseases.
    - When symptom expression is only during one phase of the plants growth cycle this must be monitored closely.
    - Virus symptoms are masked by protected cultivation and other techniques must be employed to detect the diseases - herbaceous indexing, ELISA and other serology, electron microscopy etc.
    - Support facilities for diagnosis must be at hand, so that transient symptoms must be clarified and quickly identified (this issue is dealt with in detail later).
    - Record keeping is particularly important with spray schedules when toxicities are confused with disease symptoms.
    - Growing media must be monitored so that deficiencies or toxicities are recognized also.
- Plants are grown to maturity so that seed or propagating material can be harvested for release.
- Once infected plants are detected in PEQ they should be immediately removed and destroyed, a pest/pathogen should not normally be kept in close proximity to healthy plants for more than one day.
- Insect and other vector control in the facility should ensure that the potential for disease spread is eliminated.

Infected plants

General conditions for the treatment of plants that are found to be infected with a disease can be found in the Australian Plant Quarantine and Inspection Services Manual (Anonymous, 1989) and have served effectively for many years. The procedure adopted is as follows:

"Diseased plants are to be destroyed as soon as the disease is detected (identified). All other varieties of the species in question in the post-entry facility are to be given an appropriate protective treatment and thoroughly checked before release from quarantine. In the case of virus infection all diseased plants are to be removed and destroyed and the remaining plants (of the consignment) are observed until a qualified virologist is satisfied that they are free from infection."

Wholesale destruction of imported lines is not practiced and is not warranted given the level of care that is possible within a quarantine facility and the generally slow rate of spread of diseases in well run facilities.

PEQ in isolated field plots

For some pathogen/plant combinations growth in closed greenhouse is not conducive to the expression of particular pathogens. For example, in the case of bacterial leaf spot of strawberries the plants do not show any symptoms of the disease until they are placed outside to be damaged by wind and other weather elements. In Australia, as a result of this observation, all important crop plants that could be infected with important bacterial pathogens must be grown for one year in the open before they can be released from quarantine.

Some crops could not be effectively propagated in screenhouses, for example in the production of hybrid maize from parent lines both male and females must be grown in the open. As an example, in Australia, although downy mildew of maize is considered an important seed-borne quarantine disease, post-entry quarantine is conducted in pots in the open. In field grown systems, rouging out affected plants and those closely adjacent to the initial detection is all that is required.

The matter of isolation of crops is important but should not be overdone. In the case of field planting it would be reasonable to isolate the imported seed crop from local crops of related genera according to the known activities of vectors. Two to five hundred metre isolation has often been used for this purpose, but the specific basis for this is often judged in terms of a very liberal estimate of the capabilities of insect vectors or the capacity of the wind to carry pollen. Comparisons between the rates of movement of pollen-borne viruses in fruit crops indicate very slow rates of spread, even though the pollen-borne nature of the pathogens is well documented.

Staffing requirements

If we examine the role of growing seeds in PEQ in order to avoid the establishment of new pests and diseases in uninfected areas, the prevalence of these diseases in the plants that result from the sown seed then we would be in a position to analyze the technical needs of this procedure. Kahn (1983) lists in order of priority the kinds of professional disciplines that he considers to be required for the functioning of a PEQ station. He lists the requirement for a virologist as highest and that of a horticulturist last after plant pathologists and entomologists. My experience shows that this is completely wrong. If a comparison is made of the anticipated workload of a Station then it becomes apparent that a skilled horticulturist is a far greater requirement than other professions (Chambers, 1986).

The inputs a horticulturist would have in the operation of a Station would be as follows:

- treatment of the seed and pre-conditioning recalcitrant seeds,
- sowing of seeds in an appropriate medium for seedlings, and transplanting into a suitable new medium through to development of the seed crop,
- determination of conditions for optimal growth of the plants,
- diagnosis of deficiencies/toxicities and their correction,
- determination of specific needs for propagation of difficult plants, and
- harvesting and propagation of material/seeds at the end of the PEQ term as well as the overall first detection of abnormalities and symptoms that may require further clarification by other specialists.

These activities would have to be carried out for all the seeds and plants growing in PEQ.

The other professions are not required on a full time basis. Not all seed , or even vegetatively propagated plants, in PEQ are necessarily diseased, even when taken from areas known to be infected with a pathogen. In the case of seed-borne pathogens the rate of infection is rarely above 5%, and if seed is taken from apparently healthy plants it will be less than this. Mandatory treatment for known quarantine risks will reduce this rate further, and the selection of only apparently vigorous seed will reduce it even more. A full-time virologist is therefore not needed to look at the very few plants that result from seed-borne infection in PEQ. Nevertheless the inputs from this discipline, and others is needed but generally on a referral basis only. Therefore it is operationally and economically efficient to employ a horticulturist full-time to grow the plants in PEQ and to have other professionals available close by. It is, therefore, prudent to locate PEQ facilities close to existing research/diagnostic establishments. This is the case for all PEQ stations in Australia and throughout the Asian and Pacific region. In other areas this has not happened and difficulties have been experienced in keeping staff at the location, and the result has been inefficient operation or abandonment of the facility. Also the carriage of apparently infected plants to another location in order for diagnosis to be conducted has its own obvious high risks.

Conclusions

The problems with the import of seeds and their associated pests can be effectively addressed if in the first case an efficient mechanism is in place to assess those pests which present a quarantine risk. It is not possible to address every risk simultaneously, and in any case the movement is towards the identification of those pests that present a quarantine risk i.e., that are not present and pose a potential economic risk to the country importing the seed. This may be a new concept for some who have interpreted quarantine as being aimed at all pests. With the present economic climate worldwide it would be impossible to justify a technical commitment against all pests, if this could be achieved. The processes of protection are not complex, though they have been over-emphasized in the past. The present proposals are within the general framework of achievability and sustainability, factors which today are becoming increasingly important in agriculture at a worldwide level. Simple structures which are durable and can be effectively maintained with minimal technical inputs, but with effective control by horticulturists and diagnostic professionals are the keystones to success of this most important activity.

References

Anonymous 1989. *Plant Quarantine Manual.* Australian Quarantine and Inspection Service, Dept. Primary Industries and Energy, Canberra.

Chambers, A.G. (ed.). 1986. *Nursery Stock Manual.* Australian Quarantine and Inspection Service, Dept. Primary Industries and Energy, Canberra.

Ikin, R. 1990. The International Plant Protection Convention: its future role. *FAO Plant Protection Bulletin* **38**(3): 123-126.

Kahn, R.P. 1983. *A Model Plant Quarantine Station - Principles, Concepts and Requirements.* ASEAN PLANTI, Selangor, Malaysia: 303-330.

Sheffield, F.M.L. 1968. Closed quarantine procedures. *Ann. appl. Mycol.* **47**(1): 1-8.

Smee, L. 1975. The post entry quarantine of imported plant material into Australia. *PANS* **21**(2): 168-174.

Smee, L. and Setchell, P.J. 1973. *Post-entry Quarantine for Imported Plants.* Dept of Health, AGPS, Australia.

Pests of
Quarantine Significance

Seed-borne Fungal Pathogens of Quarantine Significance

HENRIK JØRSKOV HANSEN

Danish Government Institute of Seed Pathology for Developing Countries, Ryvangs Allé 78, DK-2900 Hellerup, Denmark

The objective of plant quarantine is to act as a safety filter for exchange of planting material including seed and grain. The inspection system should be quick and simple but at the same time efficient enough to prevent introduction of fungi which are known to destabilize and decrease yields in the country of import.

According to Kahn (1977) different types of live plant material carrying different plant quarantine risks, in order of descending risk, are

- plants with roots and associated soil
- plants or cuttings with roots without soil
- dormant scions or cuttings without roots
- seeds without fruit pulp or debris (pure seed).

Although seed is considered of relatively low quarantine risk, some of the pathogens have specialized to an extent that they establish themselves and stay dormant closely associated with the seed until the time the host is susceptible to infection. Hundreds of fungi are reported seed-borne (Richardson, 1990). In wheat alone are reported seed-borne fungi belonging to more than 25 genera. Likewise, seed-borne fungal species belonging to 32 genera are reported in sorghum and 27 in soybean. The potential efficiency of seed-borne inoculum in spreading a disease is high. In one ha of wheat are distributed about 4.5 million seeds. One per cent infection level and even distribution of the infected seeds results in 4-5 infected seeds or infection loci in each square metre of the field.

Under normal conditions it is impossible to produce seed free of all fungal infections. Fortunately, most of the fungal infections have little or no quarantine significance because they have general distribution, cause low damage and are of low epidemic potential.

A choice has to be made on which of the many seed-borne fungal species should be considered as quarantine objects. As long as a fungus is included in the quarantine list of a country, it should be given due attention. Implementation of regulations will take resources either directly from the plant quarantine station itself or through payment for third country quarantine services rendered.

The difficult task of identifying the specific fungal seed-borne pathogens which should be included in the plant quarantine regulations has to be completed in the interest of protecting and improving regional (either national or international) plant production in an economic way.

Most countries are associated in regional organisations for quarantine like the Inter-African Phytosanitary Commission (IAPSC), and European and Mediterranean Plant Protection Organization (EPPO). Each country in a region has its own list of pathogens which are considered so dangerous to the domestic plant production, that efforts are made to avoid their introduction whether they are absent or already present within the country. On a regional level, those plant pathogens absent in all member countries are grouped in the so called quarantine list A1 while those present in one or few of the member countries are grouped in list A2 (Mathys and Smith, 1984). For the pathogens in list A1 an import zero tolerance is demanded by all member countries. For the pathogens contained in list A2, some countries may demand zero tolerance, but in general, the pathogens in this group may be accepted present below a specified tolerance limit.

Decisions on which pathogens should be included in the A1 list, depends on the reliability of compilation of information on the presence and distribution of the pathogens within a plant quarantine region. This, however, is not always available. Construction of a list, requires the input from many specialists within plant pathology and fungal taxonomy in order to select only those pathogens which are a real threat to the region.

The inevitability-of-establishment hypothesis states that all harmful organisms will eventually gain access to all regions of the world because quarantine can only delay the spread. This hypothesis does not give much scope for quarantine work. However, even a delay of introduction will give time for development or incorporation of resistance into presently cultivated varieties. Targeted breeding work may in this way add to minimize the effect of a quarantine object once it may be introduced. Because of the dynamics, lists of quarantine objects must continuously be updated.

During the workshop on Quarantine For Seed In The Near East Region, the following seed-borne fungal pathogens to wheat, barley, oats, rice, maize, chickpea, soybeans, cotton, sunflower, tomato, capsicum and cucurbits were suggested as possible quarantine organisms for the Near East Region:

Tilletia indica, T. controversa, T. caries, T. foetida, T. barclayana, Urocystis agropyri, Phaeosphaeria nodorum (Septoria nodorum), Gibberella zeae, Pyricularia oryzae, Cercospora oryzae, Stenocarpella maydis, S. macrospora, Setosphaeria turcica, Cochliobolus carbonum, Downy mildews, *Fusarium oxysporum* f.sp. *ciceri, Phaeoisariopsis griseola, Phytophtora megasperma* f.sp. *glycines, Phialophora gregata, Diaporthe phaseolorum* complex, *Peronospora manshurica, Septoria glycines, Cercospora kikuchii, Glomerella gossypii, Plasmopara halstedii, Diaporthe helianthi, Colletotrichum phomoides,* and *C. lagenarium.*

Before including them in the quarantine regulations of the region, each of the suggested pathogens must be evaluated carefully and subjected to pathogen risk analysis.

The pathogen risk analysis is basically an analysis of the epidemiological potential and geographical distribution of a pathogen. To these two criteria must be added the likelihood of the pathogen to be established in a given area, as well as an analysis of the cost involved in excluding the disease through quarantine regulations as compared to cost and effect on the environment incurred in controlling the disease in the field. Fungi which persist in soil once introduced to a new area as seed-borne inoculum may be considered of high risk. *Tilletia indica* and *Fusarium oxysporum* f.sp. *ciceri* are such examples. In the risk analysis each fungus/host/environment combination has to be evaluated separately.

The risk analysis must be revised regularly because changes in factors may make the result of a risk analysis invalid. Such changes could be shifts in cultivation practices, use of new varieties, development of new control measures or introduction of new crops.

Exclusion of the quarantine objects may be achieved by putting an embargo on the seed of a crop with which the fungus may be carried or by imposing restrictions such as demand of import permits, import from specified countries or areas within a country where the fungus does not occur, demand of phytosanitary certificate, testing of the seed consignment for the fungus in post-entry quarantine or compulsory seed treatments. The level of import restriction depends on the importance of the crop in the country concerned and on the reputation of the product in the international market.

How strict quarantine for seed should be is influenced by the agricultural contribution to the economy of the country concerned. Where a large amount of subsistence agriculture is practiced and plant protection measures cannot be afforded, as is the case in many developing countries, it may be advisable to protect the weak agricultural sector through a very strict but expensive quarantine organization. Early establishment of a disease may be fatal for the farmer. However, in such a case the distinction between certification for quarantine and certification for quality become unclear. Such quarantine arrangements may better be divided to create a strong national quality control service to introduce high quality seed to the farmers which is particularly important where lack of plant protection measures may prevail.

If accepting a need of more strict quarantine regulations in developing countries in order to protect domestic agriculture, the extended list of organisms must be followed up by use of efficient detection methods, sufficient staff and appropriate equipment in the plant quarantine stations.

Chandrashekar (1991) has discussed the need of introducing the concept of physiological races in quarantine and propose a model for a structured system of assessment. *Pyricularia oryzae* and *Ascochyta rabiei* are examples of two pathogens which are established and widespread within the area where rice and chickpea are cultivated. Each has many races against which resistance genes exist. However, all resistance genes do not occur in one variety. The speed by which local resistance breaks down is likely to be affected by the exchange of pathogen races along with planting material. This spread of races could be limited through quarantine regulation. However, it would require an embargo on the seed exempting breeding material free of infection or introduction of testing for fungal races in plant quarantine. None of the mentioned measures would be realistic in plant quarantine, particularly of developing countries. To overcome the most serious problems related to movement and introduction of fungal races, use of multilines could be an approach to reduce the effect of spreading races.

References

Chandrashekar, M., 1991. Relevance of physiological races of plant pathogens to quarantine. *EPPO Bulletin* **21**: 87-93.

Kahn, R.P., 1977. Plant quarantine: Principles, methodology, and suggested approaches, pp. 289-314. In: *Plant Health and Quarantine in International Transfer of Genetic Resources* (eds. W.P. Hewitt and L. Chiarappa). CRC Press, Cleveland, Ohio.

Mathys, G. and Smith I.M., 1984. Regional and global plant quarantine strategies with special reference to developments within EPPO. *EPPO Bulletin* **14**: 83-95.

Richardson, M.J., 1990. *An Annotated List of Seed-borne Diseases*. 4th edition. The International Seed Testing Association, Switzerland.

Seed-borne Bacterial Pathogens of Quarantine Significance

H.K. MANANDHAR* and CARMEN N. MORTENSEN**

*Central Division of Plant Pathology, Nepal Agricultural Research Council, Khumaltar, Lalitpur, Nepal; **Danish Government Institute of Seed Pathology for Developing Countries, Ryvangs Allé 78, DK-2900 Hellerup, Denmark

Introduction

It is well known that international exchange of seeds of plant germplasm can serve as an important means of introducing new species and or new strains of plant pathogenic bacteria. Several classical examples have been documented by Neergaard (1979) in his book "Seed Pathology". Introduction of *Clavibacter michiganensis* subsp. *michiganensis* into the U.K. from the U.S.A., *Pseudomonas syringae* pv. *glycinea* into Scotland from Sweden, *Xanthomonas campestris* pv. *campestris* into Portugal from France, and *X. c.* pv. *phaseoli* into New Zealands from the Netherlands are some examples. Schuster and Coyne (1971) detected several new strains of *X. c.* pv. *phaseoli* in bean seeds imported from Colombia and Uganda. These new strains were found more virulent than native isolates of the pathogen (Schuster *et al.,* 1973).

Seed-borne bacterial infection is not only responsible for the introduction of a new disease into an area, but also can cause severe epidemic at the same time. Very low infections in seed lots (e.g. one infected seed in many thousands) are good enough to cause severe disease in fields. Generally, levels of bacterial infection in seed stocks are often low, ranging from <0.01 to 1%, and 1% can be considered as high levels of inoculum for disease development (Frison *et al.,* 1990). Walker and Patel (1964) found that 30 infected seeds per hectare with halo blight pathogen, *Pseudomonas syringae* pv. *phaseolicola*, were enough to create severe epidemics. Even 1 infected seed per 16,000 could result in complete crop loss (Guthrie *et al.,* 1965). Similarly, Schaad *et al.,* (1980) found high incidence of black rot of crucifers (*X. c.* pv. *campestris*) with 3 or more diseased seedlings per 10,000 seeds sown.

Seed-borne bacterial pathogens

Many important plant pathogenic bacteria are seed-borne. They may also be carried by dust, crop debris or soil mixed with seeds. For most bacteria, seed-borne

*This paper replaces Dr J. Taylor's paper presented at the Workshop.

inoculum is of major importance to their survival and dissemination. They may survive as long as seed survives. A survey, based on Annotated List of Seed-borne Diseases (Richardson, 1990), revealed 8 genera, 31 species, 6 subspecies and 61 pathovars of seed-borne bacteria (Table 1). This survey includes only those bacteria which are accepted under the International Code or Standards (Bradbury, 1986; Goto, 1992).

Table 1. Seed-borne pathogenic bacteria

Genera	Number of species	Number of subspecies	Number of pathovars
Bacillus	4		1
Clavibacter	4	4	
Curtobacterium	1		2
Erwinia	5	2	1
Pseudomonas	12		23
Rhodococcus	1		
Spiroplasma	1		
Xanthomonas	3		34

Bacterial pathogens of quarantine significance

Kahn (1991) defined a pathogen of quarantine significance as a pathogen species that does not occur in a given country or an exotic strain of domestic species to that country if the pathogen is known to cause economic damage elsewhere, or has a life cycle or host/pathogen interaction that shows a potential to cause economic damage under favourable host, inoculum, and environmental conditions. He further added that an importation of a pathogen that already occurs in a given country is also of quarantine significance if an on-going regional or national containment, suppression, or eradication programme is directed against that pathogen species.

Neergaard (1979) categorized seed-borne pathogens of quarantine significance into two groups: A and B. Category A includes dangerous pathogens which do not occur in the importing country and when introduced, may spread very quickly. Category B includes those pathogens which do not occur or have a limited distribution in the importing country, and have a moderate rate of increase. On this basis of potential hazard in the international quarantine, Goto (1992) listed some examples of seed-borne bacteria under category A and B (Table 2). For category A there must be complete prohibition from areas where the disease occurs. For category B seed may be tested based on adequate sampling and the tested samples must be found completely free of infection or contamination.

The European and Mediterranean Plant Protection Organization (EPPO) has defined and listed two types of quarantine organisms: A1 and A2. The A1 includes pests not present in the region and require a common phytosanitary strategy for their exclusion and the A2 includes those which are present in the region, but subject to international phytosanitary measures to prevent their further spread (Smith, 1984). The seed-borne bacteria included in the EPPO lists are shown in Table 2.

Schaad (1988) listed 11 seed-borne bacteria (one of them, *Xanthomonas campestris* pv. *phaseoli* var. *fuscans*, is no longer accepted as different from *X. c.* pv. *phaseoli*) which are regulated by five or more countries (Table 2).

Table 2. Seed-borne bacteria of quarantine significance

Pathogen	Host	Regulating countries (no.)[1]	Quarantine status
Clavibacter michiganensis			
subsp. *insidiosus*	*Medicago*	16	A2 for EPPO[2]
subsp. *michiganensis*	*Lycopersicon*	>20	A2 for EPPO[2]
Clavibacter rathayi	*Dactylis*		Cat. A[3]
Curtobacterium flaccumfaciens			
pv. *flaccumfaciens*	*Glycine, Phaseolus*	17	A2 for EPPO[2]; Cat. A[3]
Erwinia stewartii	*Zea*	14	A2 for EPPO[2]
Pseudomonas syringae			
pv. *glycinea*	*Glycine*		A2 for EPPO[2]; Cat. B[3]
pv. *lachrymans*	*Cucumis*	5	
pv. *phaseolicola*	*Phaseolus*	10	
pv. *pisi*	*Pisum*	>17	A2 for EPPO[2]
Xanthomonas campestris			
pv. *campestris*	*Brassica*	7	Cat. B[3]
pv. *papavericola*	*Papaver*		Cat. A[3]
pv. *phaseoli*	*Phaseolus*	10	A2 for EPPO[2]
pv. *sesami*	*Sesamum*		Cat. A[3]
pv. *vesicatoria*	*Capsicum*	8	Cat. B[3]
Xanthomonas oryzae			
pv. *oryzae*	*Oryzae*		A1 for EPPO[2]
pv. *oryzicola*	*Oryzae*		A1 for EPPO[2]

[1]Schaad (1988); [2]Smith (1984); [3]Goto (1992)

Kahn (1989), in his book "Plant Protection and Quarantine," also listed some bacteria of quarantine significance. Of them which are known to be seed-borne and not included in Table 2 are:

Clavibacter michiganensis subsp. *sepedonicus* on *Solanum, C. tritici* on *Triticum, Curtobacterium flaccumfaciens* pv. *flaccumfaciens* on *Beta, Erwinia carotovora* subsp. *atroseptica* on *Solanum, E. carotovora* subsp. *carotovora* on *Nicotiana* and *Trifolium, Pseudomonas solanacearum* on *Capsicum* and *Lycopersicon, Pseudomonas syringae* pv. *atrofaciens* on *Hordeum and Triticum, P. syringae* pv. *siringae* on several plant species, *P. syringae* pv. *tabaci* on *Glycine* and

Nicotiana, Xanthomonas campestris pv. *citri* on *citrus*, and *Xanthomonas manihotis* on *Manihot*.

Criteria for determining seed-borne pathogens of quarantine significance

For determining seed-borne pathogens of quarantine significance a set of 16 criteria was developed by a group of scientists of the Biological Assessment Support Staff, PPQ, U.S. Department of Agriculture, U.S.A. (Kahn, 1989). They were later ranked by consensus of 149 respondents and are reproduced below:

1. Economic damage (high to low, many crops to one crop)
2. Occurrence in the U.S. (not known to occur or occurs)
3. Host range (wide to narrow)
4. Ease of spread from seedlings infected by seed-borne inoculum (easy to difficult)
5. Ease of detection in a sample inspected by quarantine officers using methods and equipment available at inspection stations (easy to difficult)
6. Longevity, survival in the absence of a host, dormancy (long to short)
7. Ease of establishment in the soil as a result of planting infected or contaminated seed (easy to difficult)
8. Seed-borne in other hosts that are also imported as seeds (no other hosts to many hosts)
9. Ease of spread from a planted annual crop to an established perennial crop or weeds (easy to difficult)
10. Intended use of imported seed (nonpropagation or propagation)
11. Exotic strains of a domestic species - a species occurs in the U.S., but exotic strains occurs elsewhere (none to many exotic strains)
12. Percentage of transmission through seed (low to high)
13. Probability that the pathogen could also enter along other man-made or natural pathways (high to low)
14. Volume of seeds to be imported (low to high)
15. Ease of detection and/or eradication, should a seed-borne pathogen enter on seed and become established (easy to difficult or not possible)
16. If for non-propagation, e.g., consumption or manufacture, ease of spread from the site or premise (easy to difficult)

All criteria mentioned above are common for all kinds of seed-borne pathogens, including bacteria. However, some criteria such as 4, 5 and 7 are particularly important for seed-borne bacteria. Because, spread of disease from seedlings infected by seed-borne inoculum is very high. Most inspection stations are not equipped for isolating, culturing, and characterizing bacteria present on imported articles (Kahn, 1989). And some seed-borne bacteria are soil-borne as well. They are, for examples, *Clavibacter* spp., *Curtobacterium* spp., *Erwinia stewartii, Pseudomonas solanacearum,* and several pathovars of *P. syringae*. Once they get established in uninfested soil through seed, it is very difficult to eradicate them.

Discussion

A bacterial pathogen of quarantine significance to a country or region may not be necessarily the same to another country or region. It depends on climate, agriculture and policy of a particular country or region. It is also possible that a bacterial pathogen which is not important at present might become a quarantine object in future because of changes in the agricultural practices, especially creating favourable conditions for the pathogen.

Usually when a new disease is found in an area it is claimed that the disease is introduced, most probably through seed. It is, however, not true all the time. Because a disease might have been prevalent in a country or region without notice and it becomes problem when favourable conditions are available, especially in presence of newly introduced susceptible varieties of a crop. This can be illustrated by epidemics of bacterial blight of rice (*Xanthomonas oryzae* pv. *oryzae*) in many countries of Southeast and South Asia with the introduction of IR 8, a highly susceptible rice cultivar to the disease, from IRRI. The disease was previously existed in the region (Goto, 1992).

It has already been mentioned that rather low levels of seed-borne inoculum of some bacterial pathogens are enough to cause severe diseases in fields under favourable conditions. Such information is lacking for many bacterial diseases. There are only few seed-borne bacterial pathogens for which inoculum thresholds have been establised, e.g. *Pseudomonas syringae* pv. *phaseolicola* (Taylor *et al.,* 1979), *Xanthomonas campestris* pv. *campestris* (Schaad *et al.,* 1980), *X. campestris* pv. *translucens* (Schaad and Forster, 1993).

The need for establishing inoculum thresholds is now greater than ever before because of the increasing seed movement and the need for reasonable phytosanitary requirements (Kuan, 1988). Schaad (1988) said that inoculum threshold should be established before establishing tolerance levels for planting or quarantine. He asked a question with an example that if seed is to be sown in a climate where the disease is not expressed because of cool temperatures, what is the value of a zero tolerance? The question is very practical and convincing. But the problem is that the seed produced under such conditions may go to other areas where the conditions are favourable and it is possible that an infected seed lot grown under unfavourable conditions may not always produce healthy seed. One can say that it is not a problem if the crop is not grown for seed purposes.

References

Bradbury, J.F. 1986. *Guide to Plant Pathogenic Bacteria.* CAB International Mycological Institute. 332 pp.

Frison, E.A., Bos, L., Hamilton, R.I., Mathur, S.B., and Taylor, J.D. (eds). 1990. *FAO/IBPGR Technical Guidelines for the Safe Movement of Legume Germplasm.* Food and Agriculture Organization of the United Nations, Rome/International Board for Plant Genetic Resources, Rome. 88 pp.

Goto, M. 1992. *Fundamentals of Bacterial Plant Pathology*. Academic Press, Inc. 342 pp.

Guthrie, J.W., Huber, D.M., and Fenwick, H.S. 1965. Serological detection of halo blight. *Plant Disease Reporter* **49:** 297-299.

Kahn, R.P. 1989. *Plant Protection and Quarantine*. Vol. I. Boca Raton, FL:CRC Press. 226 pp.

Kahn, R.P. 1991. Exclusion as a plant disease control strategy. *Annu. Rev. Phytopathol.* **29:** 219-46.

Kuan, T.L. 1988. Inoculum thresholds of seedborne pathogens. *Phytopathology* **78:** 867-868.

Neergaard, P. 1979. *Seed Pathology*. Vol I and II. The Macmillan Press Ltd., London and Basingstoke. 1191 pp.

Richardson, M.J. 1990. *An Annotated List of Seed-borne Diseases*. The International Seed Testing Association. Zürich, Switzerland.

Schaad, N.W. 1988. Bacteria. Part of Inoculum thresholds of seedborne pathogens Symposium. *Phytopathology* **78:** 872-875.

Schaad, N.W. and Forster, R. 1993. Black chaff (*Xanthomonas campestris* pv. *translucens*). In: *Seed-borne Diseases and Seed Health Testing of Wheat* (eds. S.B. Mathur and B.M. Cunfer). Danish Government Institute of Seed pathology for Developing Countries, Copenhagen, Denmark.

Schaad, N.W., Sitterly, W.R., and Humaydan, H. 1980. Relationship of incidence of seedborne *Xanthomonas campestris* to black rot of crucifers. *Plant Disease* **64:** 91-92.

Schuster, M.L., and Coyne, D.P. 1971. New virulent strains of *Xanthomonas phaseoli. Plant Disease Reporter* **55:** 505-506.

Schuster, M.L., Coyne, D.P., and Hoff, B. 1973. Comparative virulence of *Xanthomonas phaseoli* strains from Uganda, Columbia, and Nebraska. *Plant Disease Reporter* **57:** 74-75.

Smith, I.M. 1984. Activities of the European and Mediterranean Plant Protection Organization in relation to seed-borne pathogens. *Seed Science and Technology* **12:** 57-58.

Taylor, J.D., Phelps, K. and Dudly, C.L. 1979. Epidemiology and strategy for the control of halo-blight of beans. *Ann. Appl. Biol.* **93:** 167-172.

Walker, J.C., and Patel, P.N. 1964. Splash dispersal and wind as factors in epidemiology of halo blight of bean. *Phytopathology* **54:** 140-141.

Plant Viruses as Pests of Quarantine Significance in Seed

L. BOS* and K.M. MAKKOUK**

*Research Institute for Plant Protection (IPO-DLO), P.O.Box 9060, NL-6700 GW Wageningen, The Netherlands; ** International Center for Agricultural Research in the Dry Areas (ICARDA), P.O.Box 5466, Aleppo, Syria.

Introduction

There is a growing demand for improved cultivars to increase crop productivity, essential for the ever increasing world population. Breeding such cultivars requires international transfer of germplasm, and this may facilitate the long distance dissemination of seed-borne viruses, some of which are of quarantine significance. In order to maintain high crop productivity in modern agriculture, it is essential to minimize the risk of introducing seed-borne viruses, especially to geographic locations known to be free of such viruses.

Propagation material, some terms

In botany, any form of plant propagation material, either spore, seed, fruit or other portion of a plant, when being dispersed and able to produce a new plant, is called a **propagule** or diaspore. The term covers both vegetatively produced materials, such as **bulbs, tubers, scions** for rooting or grafting, **budwood,** and **seed,** resulting from sexual reproduction. Vegetative offspring obtained from a single plant is usually genetically homogenous; it is called a **clone**. Offspring obtained via seed is genetically variable. The term seed is sometimes confusingly being used for vegetative propagation material, as in the case of 'seed' potatoes. Seed resulting from sexual reproduction may then for clarity be called 'true seed'.

In agricultural terms, plant propagation material that is sold and distributed to growers is called **commercial plant propagation material.** It derives from **nuclear**, **basic** or **foundation stock** that, in turn, originates from **breeders' lines** that proved agriculturally valuable at extensive testing under growers' conditions. The parental material used for crossing increasingly comes from gene banks containing wide ranges of genotypes of cultivated and wild origin. Gene banks are meant for long-term preservation of the world's **genetic resources** and to prevent genetic erosion. Such material used for preservation and breeding is often called **germplasm.** The term refers to any plant propagation material with emphasis on its genetic contents.

Interregional and international transfer of seed

The interregional and international transfer of plant propagation material is increasingly playing an important role in agricultural modernization. There are three types of material that are being transferred:

Category 1. Established alien crop genotypes, introduced in small amounts into a new region or environment.

Category 2. Largely diverse genetic material, either from cultivated or wild origin, as parental material for crossing, as promising material for multilocational testing (breeders' lines), and as entries of genebanks or germplasm collections.

Category 3. Commercial plant propagation material produced by specialized growers and in special regions or countries and subsequent distribution by specialized seed traders better to exploit local growing conditions and expertise.

Category 1 is as old as international traffic and Western colonization of the Southern Hemisphere. Most world crops, such as potato, rubber, cassava and cocoa, are of alien origin. In many countries, demand for new crops is high. Category 3 has already achieved tremendous proportions. Various developing countries strive for their share in producing propagation material for the world market or are anxious to obtain such material as planting or foundation stock rapidly to improve cropping in their own countries. The movement of genetic materials for breeding purposes (Category 2) has enormously increased over the last two or three decades. Modern gene banks contain thousands of entries from all over the world. Distribution of seed from collections also involves thousands of samples annually (Bos, 1989a).

Health aspects

When alien genotypes are introduced into a new environment and unknown diseases show up on them, it is usually the seed or planting stock that is being blamed for having introduced exotic pests. However, there is a wealth of instances where local pests or pest genotypes that were previously unimportant or were hidden in the wild vegetation suddenly got a chance to emerge on alien crop genotypes that were never previously attacked by such pathogens and therefore had no chance to develop resistance or tolerance to them. Striking examples are cocoa swollen shoot, maize streak and rice yellow mottle viruses in Africa and rice tungro virus in South-East Asia. These viruses were later found to be widely present, but often without symptoms, in local crop genotypes and wild vegetation.

In many cases, however, pests are carried by germplasm. This explains our concern about germplasm health. Viruses play a special role (Bos, 1978, 1989).

(1) They are automatically transferred in vegetative plant propagules when these originate from infected plants and are increasingly found capable of doing so in seeds. Over 120 plant viruses are already known to be able to pass to plant offspring via seeds, although rates of transmission via seed are often low, or so low that transmission of the particular virus via seed escapes attention (Bos, 1977).

(2) As a rule, they remain infective as long as the seed remains viable and thus survive long-term storage and long-distance transport.

(3) After planting, they mostly spread rapidly from infected seedlings, which then act as effective sources of infection within the crop, and may cause epidemic attack of the ensuing crop.

(4) Soil-borne viruses with high chances of seed transmission spread so slowly that, by the time they attract attention, their origin may be untraceable.

(5) They often occur without symptoms, especially in dormant plant material.

(6) They are usually hard to detect in such material.

(7) As a rule, they cannot be removed by disinfection.

The risks of viruses in plant germplasm are twofold. First comes the direct effect on the ensuing plants and crops, and this is of special concern with commercial stock. Many seed-borne legume viruses, such as those of blackeye cowpea mosaic, pea seed-borne mosaic, peanut stripe and soybean mosaic, are spread by aphids. In crops partially infested from seed, their incidence may increase up to 100% within two months, depending upon population density and activity of the vector. Hence economic damage may arise rapidly. Several of these viruses have already achieved world-wide distribution through their propagule-borne nature. Major concern is then to avoid economic damage or to keep it below a certain threshold. Some degree of infection of the propagation material is then tolerated, because absolute freedom would be prohibitively expensive. The tolerance may differ between recipient countries. The relative absence of viruses in seed is then a matter of **seed quality.**

Those viruses that are potentially harmful and still of limited distribution should be prevented from spreading to other countries or regions and they should receive **quarantine status.** In the developing countries, often in demonstration fields of new genotypes, virus infection is observed right from the beginning. This infection must obviously have originated from seed-borne inoculum. Introduction with germplasm often attracts little attention. Several seed-borne viruses are spread within the soil by nematodes (e.g. tobacco ringspot virus causing bud blight in soybean) and some by soil-inhabiting fungi (peanut clump virus). Spatial spread by nematodes is slow and it may take ten years to achieve 50% incidence. However such viruses often also infect different crop species and various weeds without showing symptoms. In them, rates of seed transmission may also be high. So at the new site, the presence of such viruses may attract attention only long after introduction. But once the virus gets established there, it will remain for ever.

Mechanisms of seed transmission

It is not known why certain viruses are seed-transmitted and others not. Viruses that go systemic in their host plants can enter the ovule via the vascular bundle. Whether they are able to pass via the seed to the offspring depends on a number of factors (Bos, 1989b).

Those viruses that are limited to the phloem, and are transmitted in a persistent manner by insects which feed upon the phloem contents,are not seed-transmitted. They cannot get into the embryo of the seed. The same holds for mycoplasmas which, in plants, behave like phloem-limited viruses.

There are two ways of virus transmission through seed. One way is by a small group of viruses, such as tobacco mosaic, tomato mosaic and cucumber green mottle mosaic viruses. In spite of high infectivity and stability, they cannot enter the embryo. Seed infection is then limited to the integuments and the nucellar tissues, which later develop into the seed coat. These viruses remain infective there after maturation of the seed. It is not known why such stable viruses cannot get into the embryo. They may also externally contaminate the seed in fruit pulp remnants. Such viruses can be removed by heat and chemical treatments. They can only pass from the seed coat into the seedling when this is handled at transplanting.

Viruses that can reach the embryo can only do so if infection takes place before the egg cell is fertilized, or, more simply, before pollination. Thereafter, contact between embryo and mother plant through plasmodesmata is disrupted; of course, there is no vascular contact between embryo and mother plant. Such viruses can also get into an embryo on a virus-free mother plant via pollen.

Most seed-transmitted viruses are so in the embryo. In immature seed they can also be detected in the seed coat, but these viruses are less stable than tobacco mosaic virus, tomato mosaic virus and cucumber green mottle mosaic virus, and lose infectivity when the seed matures. When intact seeds are tested with highly sensitive serological methods, virus protein may be detected but this does not prove infectivity. The test then yields 'false positives'. Once in the embryo, viruses remain there as long as the seeds remain viable, as during long-term storage, e.g. in gene banks.

Seed-coat mottling, such as that of soybean seeds, may indicate that such seed originated from a mother plant infected with seed-transmitted soybean mosaic virus. Then, the mottling merely suggests the chance of seed transmission. The virus in the seed coat loses infectivity during seed maturation and there is no risk of it passing from the seed coat into the embryo or seedling.

The possibility and extent of seed transmission differ according to virus and strain, and depend highly on the host species and cultivar; temperature may also play a role.

Quarantine status

How can we judge whether a virus or other pests is of importance for quality only or should at all cost be kept out of a country or region? The FAO International Plant Protection Convention defines a quarantine pests as "a pest of potential national economic importance to the country endangered thereby and not yet present there, or present but not widely distributed and being actively controlled". How is potential economic importance assessed, especially if experience in crop loss assessment is lacking? Losses may not become apparent until a new crop genotype with susceptibility or sensitivity reveals its presence. If there is no apparent direct or potential economic effect, the above definition does not justify taking quarantine measures.

In brief, the following factors are worth considering in analyzing pest risks:

(1) Extent of distribution of the pest in and outside the country
(2) Risk of establishment in the country or region, depending upon the
 - availability of susceptible hosts including weeds
 - possibility of spread to such hosts
(3) Risk of epidemic build-up depending upon
 - number of over-wintering sources
 - vector efficiency and population density
 - crop susceptibility
(4) Risk of ensuing economic damage (actual or potential) depending upon
 - crop sensitivity

Thus, decisions whether a certain pest should be kept out are hard to take. Aspects 2 to 4 depend heavily on variable ecological factors and on the crops and cultivars grown. With respect to Item 1, it should be stressed that many countries are trying hard to keep out pests supposed not yet to occur in their country, ignoring the fact that domestic surveying for pests is scanty, to say the least. Hence final decisions are often political rather than technical. In relationships between industrialized countries, decisions about which pests should be included in official lists of quarantine organisms are often influenced by commercial interests.

Lists of viruses of quarantine significance

Most countries issue a list of viruses of quarantine importance (Holdeman, 1986) and such lists are up-dated on a regular basis. On a world-wide scale, 280 viruses were listed in the quarantine regulations of various countries. Some of these viruses may be covered by the term "virus-like agents" in some regulations. Lists which cover viruses of quarantine significance to specific regions such as North America

(Schoulities *et al.*, 1983), European and Mediterranean region (EPPO, 1986) are also published.

In developing countries, and in view of the often blatant lack of information on viruses that are already occurring, it is not easy to judge which viruses should be considered of quarantine significance. In such cases extensive surveys to identify what are the seed-borne viruses that already exist in a specific country or region is a prerequisite for the establishment of a realistic list of seed-borne viruses of quarantine importance. For example, an early survey of faba bean crop in West Asia and North Africa has shown a number of seed-borne viruses already to be prevalent is several of them (Makkouk *et al.*, 1988).

Realistic quarantine

International bodies involved in agricultural development (FAO, World Bank, IARCs, etc.) emphasized in recent discussions the importance of strengthening quarantine (Hewitt and Chiarappa, 1977; Kahn, 1977; Plucknett and Smith, 1988). Economy of countries which depend on few crops can benefit from plant quarantine. However, it is only possible to test incoming material for known viruses, and in many countries information, as mentioned above, is still limited. Furthermore, to ensure appropriate quarantine decisions, facilities and expertise need to be significantly improved in many developing countries. Despite such improvements, fool-proof systems may remain unattainable. There is a strong tendency to liberalize trade and traffic of commercial plant propagation materials and there are several other loopholes for viruses to move across geographic and natural borders. Moreover, high-security quarantine facilities are often vulnerable in developing countries.

Differences of opinion between quarantine personnel and breeders continue. The former tend to be conservative whereas the latter are more liberal, and tend to ignore quarantine regulations. It is generally accepted that we need more productive, better and more resistant cultivars, and new crops are sought. As a result, there is a pressing need for germplasm movement for further crop improvement. Thus, approaches should be realistic and aim for "improvement of the rapid and safe global transfer of germplasm" (Chiarappa and Karpati, 1980). This requires compromise and acceptance of certain risks. However, we must remain aware of such risks and try to reduce them to a minimum.

References

Bos, L. 1977. Seed-borne viruses. pp. 39-69. In: *Plant Health and Quarantine in International Transfer of Genetic Resources*. (eds. W.B. Hewitt and L. Chiarappa). CRC Press, Cleveland, Ohio.

Bos, L. 1989a. Germplasm health and international crop improvement, with special reference to viruses. pp. 19-29. In: *Introduction of Germplasm and Plant Quarantine Procedures*. (eds. A.H. Jalil *et al.,*). Asean Plant Quarantine and Training Institute, Malaysia.

Bos, L. 1989b. Virus transmission via seed: mechanisms, detection, and implications for quality and quarantine. pp. 115-118. In: *Introduction of Germplasm and Plant Quarantine Procedures*. (eds. A.H. Jalil *et al.*). Asean Plant Quarantine and Training Institute, Malaysia.

Chiarappa, L. and Karpati J.F. 1980. Safe and rapid transfer of plant genetic resources: a proposal for a global system. Paper presented at the *FAO/UNDP/IBPGR Meeting on Crop Genetic Resources*. Report FAO, Rome. 10 pp.

EPPO (European and Mediterranean Plant Protection Organization). 1986. *Data Sheets on Quarantine Organisms*. Data Sheets 1-159. OEPP/EPPO, Paris, France.

Hewitt, W.B. and Chiarappa, L. (eds). 1977. *Plant Health and Quarantine in International Transfer of Genetic Resources*. CRC Press, Cleveland, Ohio. 346 pp.

Holdeman, Q. (ed). 1986. *Plant Pests of Phytosanitary Significance to Importing Countries and States*. 5th edition. California Department of Food and Agriculture, Sacramento.

Kahn, R.P. 1977. Plant quarantine: principles, methodology and suggested approaches. pp. 289-307. In: *Plant Health and Quarantine in International Transfer of Genetic Resources*. (eds. W.B. Hewitt and L. Chiarappa). CRC Press, Cleveland, Ohio.

Makkouk, K.M., Bos, L., Azzam, O.I., Koumari, S. and Rizkallah, A. 1988. Survey of viruses affecting faba bean in six Arab countries. *Arab Journal of Plant Protection* **6**: 53-61.

Plucknett, D.L. and Smith, N.J.H. 1988. Plant quarantine and the international transfer of germplasm. *CGIAR Study Paper* NO. **52**. World Bank, Washington D.C. 52 pp.

Schoulities, C.A., Seymour C.P. and Miller, J.W. 1983. Where are the exotic disease threats? In: *Exotic Plant Pests and North America Agriculture*. (eds. C.L. Wilson and C.L. Graham). Academic Press, New York.

Nematodes as Pests of Quarantine Significance in Seed

GEORGES CAUBEL

Institut National Recherche Agronomique Zoologie, B.P. 29, 35650 Le Rheu, France

Introduction

Plant-parasitic nematodes have proved to be important economically all over the world. Many species have a rather wide distribution, others may be restricted to small areas. A great number of nematodes present in soil, in plants or in seed, are dispersed by farming implements, movement of commercial consignments from one country to another as well as exchange of germplasm across international boundaries.

Seed-borne nematode species

Some nematode species may be occasionally dispersed with seeds, e.g. juveniles of *Meloidogyne* sp. in lucerne (*Medicago sativa*), live juveniles of *Heterodera glycines* and *H. schachtii* as cysts with seeds of soybean (*Glycine max*) and sugar beet (*Beta vulgaris*), respectively. Dispersal of cysts is uncommon, but is of very great importance for the introduction of a species into a new area.

Many records are found of true seed transmission in literature, even in important crops (Table 1).

On rice (*Oryza sativa*), the white-tip disease, caused by *Aphelenchoides besseyi* was found in a great number of rice growing areas of Asia, tropical America and Africa (Fortuner and Orton Williams, 1975). Yield losses incurred by this pest amount to as much as 30% to 50%, depending on climatic and cultural conditions. This is the most important nematode of rice. This species is parthenogenetic, prolific as numerous generations occur in the plant. Other hosts are *Pennisetum*, *Panicum*, strawberry and fungi. *Aphelenchoides* which are capable of withstanding desiccation may be found in a quiescent state beneath the hulls of rice grain. When infested seed is sown, the nematodes become active and move to the growing point of stem and leaves of seedlings. They feed ectoparasitically and provoke crinkling and distortion of the flag leaf enclosing the panicle. The affected plants produce panicles reduced in size and number. Nematodes become quiescent upon maturation of grain.

Another important pest of rice, *Ditylenchus angustus*, causes the "Ufra" or "dak-pora" disease. This obligate and highly specialized ectoparasite of rice, in spite of being found inside the glumes of the panicle, seems to be unable to spread through seeds (Hashioka, 1963). Losses are very high. Nematodes feed ectoparasitically causing malformations of host tissues; infected plants are stunted with grains often having lesions and leaves often wilt.

The genus *Anguina* is recorded on Graminaceae. Seed infestation of various grasses by *A. agrostis* (considered by some to be *A. funesta*) is essential as this obligate parasite does not persist for long in soil. Galls formed on *Agrostis* instead of grain are toxic to livestock.

Table 1. List of nematodes transmitted by seed.

Nematode	Crop	Seed species	Distribution
Aphelenchoides besseyi	Rice	*Oryza sativa*	Asia, America, Africa
A. ritzemabosi	Aster	*Callistephus sinensis*	Europe
A. blastophthorus	Aster	*Callistephus sinensis*	
Anguina tritici	Wheat, rye	*Triticum aestivum*, *Secale cereale*	Europe, Asia
A. agrostis	Bentgrass	*Agrostis* spp., *Lolium* spp.	Europe, United States, Australia
A. funesta	Rye grass	*Lolium rigidum*	Australia
Subanguina chrysopogoni	Grass	*Chrysopogon fubus*	Asia
Ditylenchus angustus	Rice	*Oryza sativa*	Asia, Egypt
D. dipsaci	Oat	*Avena sativa*	Europe
	Onion	*Allium cepa*	Europe
	Shallot	*Allium* spp.	Europe
	Beet	*Beta vulgaris*	Europe
	Fuller's teasel	*Dipsacus fullonum*	Europe, America
	Cat's ear	*Hypochaeris radicata*	America
	Lucerne	*Medicago sativa*	Europe, New Zealand
	Plantain	*Plantago major*	America
	Dandelion	*Taraxacum officinale*	America
	Clover	*Trifolium* spp.	Europe
	Field bean		
	Broad bean	*Vicia faba*	Europe, Africa, Middle East
	Carrot	*Daucus carota*	Europe
	Runner bean	*Phaseolus* spp.	Europe
	Pea	*Pisum sativum*	Europe
	Buckwheat	*Fagopyrum sagittatum*	Europe
	Spring vetch	*Vicia sativa*	Europe
D. destructor	Groundnut	*Arachis hypogea*	Africa
Heterodera schachtii	Beet	*Beta vulgaris*	Europe
Panagrolaimus sp.	Pearl millet	*Pennisetum americanum*	Asia
Rhadinaphelenchus cocophilus	Coconut	*Cocos nucifera*	Tropical America

A. tritici, a typical seed-borne nematode, induces formation of galls (ear cockles), which are easily recognized by their small size and brown colour. The galls, full of anabiotic juveniles, are a major source of dissemination and infection. *A. tritici* remains a problem in many countries and may limit trade exchanges, for this species is entirely dependent on seed dispersal (Parruthi and Bhati, 1981).

The genus *Panagrolaimus* is recorded in seed of pearl millet (*Pennisetum americanum*). In India, Panchbhai and Varma (1986) consider this seed-borne nematode as a significant plant quarantine object.

The stem nematode, *Ditylenchus dipsaci*, is a very common and poly-phagous species. It is able to contaminate seeds of important crops (Richardson, 1979). Dissemination is well known in lucerne and red clover. Bingefors (1967) described its spread to different countries. Infection is also known in spring vetch (*Vicia sativa*) and white clover (*Trifolium repens*). On onion (*Allium cepa*), the stem nematode is a very noxious pest and dispersal by seed is a problem. With other vegetables, Green (1979) believes that a risk exists of introducing stem nematode into new areas, even with a small amount of infestation. On *Vicia faba* bean, two races occur, the giant race and the oat/onion one. The giant race is a serious pest on broad beans in Mediterranean countries, but it is also reported from England, Germany and France (Caubel *et al.*, 1972). If the infestation is severe, necrotic patches containing the nematodes can be observed by removing the testa; sometimes *D. dipsaci* are massed in clusters to form "eelworm wool". The consequences of the use of infested seed lots on the dissemination of this nematode are important. There are more nematodes in the plants if the inoculum is seed-borne. After harvest, only a small number are found in the soil; most are in the tissues of the plant, mainly in pods and seeds (Ighil and Caubel, 1986).

D. destructor has been found recently in hulls and seeds of peanut (Jones and De Waele, 1988). Infected hulls have brown necrotic tissue at the point of connection with the peg, and a black discolouration appeared along the longitudinal veins. Infested seeds were shrunken, and testae and embryos had a yellow to brown discolouration.

Juveniles of *Heterodera schachtii* have been found inside sugar beet seed. This nematode species, only found in beet and cruciferous cropping areas, may be introduced with seeds into new regions.

Infestation of commercial seed stocks is presented in Table 2. Hooper (1971) reported that up to 10% of samples of field bean submitted for certification were infested with *D. dipsaci*. However, in non-cleaned stocks of lucerne the percentage reached 14 (Caubel, 1972).

More information is needed to know the exact situation, as nematological analyses of seed are rarely undertaken on a routine basis, in comparison with other pathological tests. Transmission by seed appears to be, in some cases, the most prevalent mode of infestation of crops though, of course, soil and plant contamination also occur.

Table 2. Percentages of nematode-infested seed stocks.

Nematode	Host	% of infested seed lots	Seed lots examined
Ditylenchus dipsaci	Onion	6	685
		1	1,762
		5	
		11-27	
	Shallot	40	16
	Red clover	7	
		6	202
	Lucerne	14	496
		11	
		12	24
	Field bean	10	
		16	55
		19	136
		32	275
	Broad bean	38	58
		79	246
		53	
	Runner bean	17	42
	Pea	3	58
	Red beet	45	33
	Carrot	36	45
Aphelenchoides besseyi	Rice	70	32
		6	474
Anguina tritici	Wheat	33	103
Ditylenchus destructor	Peanut	73	877

Transmission and epidemiology

Localization in seeds

Nematodes may be present with, on or in the seed. Evidence of desiccated *D. dipsaci* adhering to the seed surface has been reported in teasel, clover and lucerne. The nematode present in the inflorescence makes the flower buds twisted and thickened. The stem nematode is present mostly in debris associated with seeds of these species (Tables 3 and 4).

In *Vicia* attacked by *D. dipsaci*, holes may be seen on the coat and necrosis on the cotyledons. On rice, *A. besseyi* penetrates the florets before anthesis and multiplies intensively; nematodes coil and become quiescent in the seeds. Most of them are present in the inner part of the husks and some remain on the kernel. *Aphelenchoides* ritzemabosi and A. blastophthorus on Callistephus are found between the seed coat and the embryo, but the embryo itself is not infected.

Table 3. Levels of contamination by *D. dipsaci* in debris and seeds. Numbers of specimens per 100 g of material.

		Over separator	Under separator	Ventilation "rice"+	"magnetic"	SEED
Red clover 1(*)	Debris %	1	3	5	17	75
	D. dispaci	1 080	90	560	360	10
Red clover 2(*)	Debris %	36	1	13	0	50
	D. dispaci	16 000	1 850	19 600	0	460
Lucerne 1	Debris %	2	2	4	2	90
	D. dispaci	4 000	560	6 250	4 480	50
Lucerne 2	Debris %	1	2	2	4	91
	D. dispaci	1 650	320	320	440	5

* Manually harvested

Table 4. Infestation levels of *D. dipsaci* after cleaning a seed stock of lucerne.

Seed purity	Number of nematodes/100 g seed
100%*	2
100%	8
99% to 99.5%	4 to 8
98% to 98.5%	26 to 106

*Separation by hand

Table 5. Maximum number of nematodes contained in seeds.

Nematodes species	Seed	Number of nematodes per unit
A. besseyi	Rice	6 4 i n
one seed		
A. tritici	Wheat	About 30,000 in one gall
D. dipsaci	Field bean	19,780 in one seed
	Broad bean	About 10,000 in one seed
	Lucerne	5 in one gram of seed
	Onion	75 in one gram of seed
Panagrolaimus sp.	Millet	84 in one seed

Nature of inoculum

All seed-borne nematodes have a great ability to survive in seeds for a long time. The larval stages in which the nematodes survive differ from one nematode to another, e.g. the second stage for *A. tritici* and the fourth for *D. dipsaci.* Generally the surviving stage is the infective stage.

The females of *A. tritici* lay eggs in young galls and the second stage larvae emerge at harvest time. Each gall contains up to 30,000 juveniles. They are known to resist desiccation for up to 28 years. *D. dipsaci* is also able to survive in the anhydrobiotic fourth stage larvae for numerous years.

The number of nematodes contained in seed is very variable (Table 5). Sometimes, the inoculum may be dispersed with plant debris, as in case of stem nematode of lucerne. In other cases, inoculum is concentrated as in *Vicia* where the number of nematodes may reach more than 10,000 in heavily infested seeds. When the nematodes induce gall formation, the number of infective juveniles may also be considerable. The quality and quantity of individuals in the stock influence the pattern of attack after sowing and thus to be considered the number of parasites on each square metre of the crop in relation to plant density (Green, 1979).

Establishment of primary infection

The nature of the nematode and its distribution determines the pattern of attack in a field. The speed of development of stem eelworm in a field is rapid and characterized by the formation of round, primary foci. Green (1979) showed the relationship between the number of foci and inoculum. In field bean crops, isolated plants infected by stem nematode are observed resulting from a low level of contamination of seeds.

In general, the ease of dissemination of nematodes with and by seeds explains the worldwide distribution of nematodes. It is generally believed that lucerne race of stem nematode has been distributed to different countries with seed from south-west Asia where lucerne is native.

References

Bingefors, S. 1967. International dispersal of nematodes. *Netherlands Journal of Plant Pathology* **73**: 44-60.

Caubel, G., Chaubet, B., Aït Ighil, M. and Marzin, H. 1972. Féverole : présence en France d'une race géante de nématode. *Phytoma* **341**: 19.

Caubel, G. 1972. Les nématodes et les semences de légumineuses. *Bulletin de la Fédération Nationale Multiplicateurs de semences*: 2-8.

Fortuner, R. and Orton Williams, K.J. 1975. Revue de la littérature sur *Aphelenchoides besseyi*, nématode causant la maladie "white tip" du riz. *Helminthological abstracts* **44**: 1-40.

Green, C.D. 1979. Aggregated distribution of *Ditylenchus dipsaci* on broad bean seeds. *Annals of applied Biology* **92**: 271-274.

Hashioka, Y. 1963. Le nématode de la tige du riz *Ditylenchus agustus* en Thaïlande. *Bulletin Phytosanitaire de la FAO* **11**: 97-102.

Hooper, D.J. 1971. Stem eelworm *Ditylenchus dipsaci*, a seed- and soil-borne pathogen of field beans (*Vicia faba*). *Plant Pathology* **20**: 25-27.

Ighil, M.A. and Caubel, G. 1986. Contamination des graines de *Vicia faba* par le nématode des tiges, *Ditylenchus dipsaci*, Conséquences épidémiologiques. *Seed Science and Technology* **14**: 431-438.

Jones, B.L. and De Waele, D. 1988. First report of *Ditylenchus destructor* in pods and seeds of peanut. *Plant Disease* **72**: 453.

Panchbhai, S.D. and Varma, B.K. 1986. Presence of *Panagrolaimis* sp. in seed of pearl millet. *Nematologica* **26**: 236-237.

Paruthi, I.J. and Bhatti, D.S. 1981. Estimation of loss in wheat yield due to *Anguina tritici* and seedgall infestation in market grains. *Indian Journal of Nematology* **11**: 15-18.

Richardson, M.J. 1979. An annotated list of seed-borne diseases. 3rd edition. Phytopathological Papers No 23. Commonwealth Mycological Institute, Kew, England.

Schreiber, E.R. 1978. Biologie, importance et moyens de contrôle du nématode des tiges sur fèves au Maroc. *Bulletin de Protection des Cultures* 4: 3-20.

Seinhorst, J.W. and Koert, J.L. 1969. Stengelaaljes op uienzaad. *Gewasbescherming* **2**: 25-31.

Insect Pests of Quarantine Significance in Seed

KLAUS RICHTER

University of Leipzig, Institute of Tropical Agriculture, Fichtestr. 28, D-7030 Leipzig, Germany

Introduction

Seeds are subject to attack by more than one hundred species of insects. An overview of selected stored pests with significance to seed is given in Table 1.

Seed infestation by storage insects leads to loss in weight and in seed quality, especially the germination ability. Even in case of a slight damage on the germ, frass symptoms appear on the cotyledons and primary leaves. According to their capacity of inflicting damage, the grain storage insects can be classified into two general groups: 1. Primary pests, and 2. Secondary pests.

Primary pests

Insect pests belonging to this group are capable of attacking healthy, mechanically undamaged and dry seed that is in best storable condition. Primary pests in relation to their specific development can be subdivided into two characteristic groups: insects that develop mainly inside the kernels, and insects that develop outside the kernels (Table 2).

Primary pests causing internal infestation

The major feeders inside the kernels of maize, sorghum, millet and rice are *Sitophilus* spp., lesser grain borer, larger grain borer, and the Angoumois grain moth.

Sitophilus spp. deposit their eggs inside the kernel, covered by gelatinous plugs (important for detection by staining), whereas all other pests of this group deposit the eggs on the surface or freely between the kernels. The newly hatched larvae tunnel into the kernels after some days. The life cycle up to the emergence of the adults takes place within the kernels, generally without easily visible symptoms. Bruchids seed beetles have a similar life cycle on legumes.

Table 1. Major insect pests of stored seed

Scientific name	Common name	Family
Major pests of stored cereals		
Trogoderma granarium	Khapra beetle	Dermestidae
Sitophilus granarius	Granary weevil	Curculionidae
Sitophilus oryzae	Rice weevil	Curculionidae
Sitophilus zeamais	Maize weevil	Curculionidae
Sitotroga cerealella	Angoumois grain moth	Gelechiidae
Rhizopertha dominica	Lesser grain borer	Bostrichidae
Ephestia elutella	Tobacco moth	Pyralidae
Ephestia cautella	Tropical warehouse moth	Pyralidae
Plodia interpunctella	Indian-meal moth	Pyralidae
Cryptolestes ferrugineus	Rusty grain beetle	Cucujidae
Cryptolestes pusillus	Flat grain beetle	Cucujidae
Oryzaephilus surinamensis	Saw-toothed grain beetle	Cucujidae
Corcyra cephelonica	Rice moth	Pyralidae
Prostephanus truncatus	Larger grain borer	Bostrichidae
Major pests of stored legumes		
Acanthoscelides obtectus	Common bean weevil	Bruchidae
Callosobruchus chinensis	Adzuki bean weevil	Bruchidae
Callosobruchus maculatus	Cowpea weevil	Bruchidae
Zabrotes subfasciatus	Brazil bean weevil	Bruchidae
Major pests of stored oilseed		
Ephestia cautella	Tropical warehouse moth	Pyralidae
Trogoderma granarium	Khapra beetle	Dermestidae
Plodia interpunctella	Indian-meal moth	Pyralidae
Oryzaephilus mecator	Merchant grain beetle	Cucujidae

Table 2. Classification of store insect pests according to their damaging potential

Primary pests		Secondary pests
With internal infestation (insects that develop mainly inside the kernels)	Without internal infestation (insects that develop outside the kernels)	Without internal infestation (insects that develop outside the kernels
Sitophilus granarius granarium *S. oryzae* *S. zeamais*	*Trogoderma granarium*	*Cryptolestes ferrugineus* *C. pusillus*
Rhizopertha dominica	*Nemapogon granellus*	*Oryzaephilus surinamensis*
Prostephanus truncatus	*Corcyra cephalonica*	*Tribolium castaneum* *T. confusum*
Sitotroga cerealella	*Plodia interpunctella*	
Species of the Bruchidae family	*Ephestia cautella*	

Table 3. The population build-up of primary and secondary insect pests on wheat (26°C ; 65-75% R.H.; 30 g of wheat)

Storage period (days)	Infestation level (No. of adults)	Mean adult population			
		Primary pests		Secondary pests	
		Sitophilus oryzae	*Rhizopertha dominica*	*Oryzaephilus surinamensis*	*Tribolium castaneum*
70	30	727.2	302.8	41.0	30.0
105	30	984.6	656.4	51.6	31.6

According to Swatonek (1975) the life cycle of *S. granarius* completes in 30 days at 26°C and 60 days at 20°C. A female of this species may deposit a total of about 400 eggs, daily 2 eggs at 26°C and 2 eggs during 3 days at 20°C (Jacobi, 1982).

In warm climates all representatives of this group are more or less good fliers and able to infest the kernels in the field before harvest. Although the adults are destroyed by the harvesting and threshing process, the kernels come to the store infested but appear healthy. Therefore it must be noted here that the group of primary pests which develops within the grain plays an extraordinary part in seed health.

The bruchids are also primary pests of seed legume seeds. Some countries e.g., Russia and countries adjoining the Black Sea, have paid high attention to pests (Spirina, 1982; Dzivilevkov, 1979). *Acanthoscelides obtectus* was the most dangerous with more than one hundred species (Spirina, 1982). Another species, *Zabrotes subfasciatus*, is a heavy pest of beans mainly in Latin America, but it is now spreading (Yus Ramos, 1976).

Primary pests without internal infestation

The most significant pest belonging to this group is *Trogoderma granarium*. It is also the only pest, included in the EPPO List A 2. This is because of the fact that larvae of *Trogoderma* are extremely insensitive to many post-harvest insecticides in comparison to insects. A detailed description of quarantine requirements of *T. granarium* is given in Data Sheet No. 21 of the EPPO List A 2.

Secondary pests

The group of secondary pests includes numerous species. All of them develop mainly outside the kernels and do not show internal infestation. Secondary pests are not able to build up populations of high density on healthy, mechanically undamaged and dry seed (Zacher and Lange, 1964; Richter, 1975). For demonstrating these differences, a comparision of both groups is shown in Table 3 (Ritcher and Gure, 1991).

Among this group many species with a high potential for mass propagation e.g., *Cryptolestes* spp., *Orycaephilus* spp., are to be found. Heavy damage to seed by these pests occurs if the grain is mechanically injured or if the stored seed has a high moisture percentage.

Ecological aspects

Fundamental investigations about climatic requirements of store pests, carried out by Howe (1965) have shown that the most insects in the store are able to produce high infestation levels at optimum climatic conditions. The estimated monthly rates of increase for some selected common pests are 12.5% for *Trogoderma granarium*, 50-60% for *Cryptolestes* spp., and 60-70% for *Tribolium* spp.

It must be stated that the mass propagation of the most store insect pests can be prevented by storage temperatures near 12°C. The distribution of insect pests in bulk grain was investigated by many workers. It is found that the highest percentage of coleopterans present in the bulk grain lives in the upper layer of about 2 m. This observation gives a good indication about sampling with simple sampling spears.

Quarantine requirements

Quarantine significance of individual organisms is based on three primary factors: (1) the value placed on the production of products infested by the pest, (2) the nature and anticipated damage and economic losses caused, and (3) the availability of technology to control or eradicate the organism.

Store pests of seed are of economic importance. Attention should, therefore, be paid to the insect pest problems. Insects or their fragments (filth) are a quality reducing factor. On the other hand it must be mentioned that phytosanitary requirements might be used as non-tariff barriers to international trade (Ikin, 1990). Therefore, trading partners must harmonize their quarantine regulations using bilateral or multilateral agreements (Hedley, 1990).

References

Collins, P.J. 1986. Genetic analysis of Fenitrotion resistance in the Sawtoothed grain beetle, *Orycaephilus surinamensis* (Coleoptera:Cucujidae). *J. of Economic Entomology* **79**: 1196-1199.

Dzivilevkov, G. 1979. Njakoi problemi pri borbata a kalifornijskata scitonosna v ska i fasulevvija rnojad. (Some problems in the control of San-Jose-Mealybug and the common bean weevil). *Rastitelna Zascita* **6**: 9-10.

Hedley, J. 1990. Improved plant quarantine using bilateral quarantine agreements for the exclusion of high-risk pests. *FAO Plant Protection Bulletin* **38**: 127-133.

Howe, R.W. 1965. A summary of estimates of optimal and minimal conditions for population increase of some stored products insects. *J. of Stored Products Res.* **1**: 177-184.

Ikin, R. 1990. The International Plant Protection Convention: its future role. *FAO Plant Protection Bulletin* **38**: 123-126.

Jacobi, M. 1982. Getreidebearbeitung und lagerung. Berlin.

Richter, K. 1975. Vorratsschutz in der Pflanzenproduktion. Leipzig. 205 pp.

Richter, K. and Gure, A. 1991. Untersuchungen über Wechselbeziehungen zwischen Primär-und Sekundärschädlingen. unpublished.

Spirina, T.S. 1982. Zernovki-opasnye vrediteli zapasov. (The bruchid seed beetles - dangerous pests on stored products). *Zascita Rast.* **9**: 40-45.

Swatonek, F. 1975. Untersuchungen zur Bilogie des Kornkäfers, *Sitophilus granarius* (L.) I. Teil. *Bodenkultur* **26**: 278-290.

Yus Ramos, R. 1976. Las especies de Brùquidos (gorgojos de las leguminosas) deinterés agricola y fitosanitario (Col.*Bruchidae*). II: Sistematica y biología. *Bol.Serv. Plagas* **2**: 161-203.

Zacher. F. and Lange, B. 1964. Vorratsschutz gegen Schädlinge. Berlin und Hamburg.

Seed Health Testing

Sampling Procedures

HENRIK JØRSKOV HANSEN

Danish Government Institute of Seed Pathology for Developing Countries, Ryvangs Allé 78, DK-2900 Hellerup, Denmark

Introduction

The main objective of the work carried out in plant quarantine laboratories is to detect the presence of pathogens which are unwanted in the examined material. The only way of making sure that no unwanted pathogen passes the plant quarantine inspection is to check every unit which is imported or exported. When dealing with seed this is obviously not possible. Since the pathogens with zero tolerance level, listed mainly in the category A_1, are considered a highly significant threat to the plant production of the importing country, even when present in traces, the quarantine inspection for such organisms should be carried out during crop production in the exporting country. This is to ensure that the disease in question is not present in the area of cultivation. Checking absence of a pathogen in representative samples is not allowed for pathogens with zero tolerance requirement.

Other pathogens of quarantine significance, for which zero tolerance level is not required but which are still unwanted in the importing country, are found in the category A_2. They are considered not to present an immediate threat to production crops in the importing country, provided only a small amount of inoculum is present in the seed. For checking consignments for pathogens listed in this category, drawing of a representative sample at the place of import is allowed. Sub-samples drawn from this sample are then used in laboratory tests.

Seed consignments may vary from a few grams of germplasm to whole shiploads. In order to draw a sample of ultimately a few hundred seeds which will be tested and the result is supposed to represent the condition of the entire consignment, the International Seed Testing Association (ISTA) has worked out specific rules stating the minimum requirements for drawing representative samples. The rules have been worked out for sampling of seed in domestic seed certification programmes, but can be followed in the plant quarantine work. If required, sampling intensity may be increased to obtain a larger seed sample than recommended for certification purposes.

A seed consignment may be of any size, but the sampling efficiency is dependent on the amount of seed in which sampling is performed. Therefore the term 'seed lot' has been introduced by ISTA.

A seed lot is defined as a portion of a consignment assumed to be reasonably uniform. The maximum size of a seed lot varies with the plant species. A few examples are presented in Table 1.

Table 1. Examples of maximum seed lot sizes. Seed consignments larger than the maximum seed lot size must be subdivided into units not larger than the maximum seed lot size (ISTA, 1985)

Plant species	Maximum size of seed lot
Very large seed (Maize)	40,000 kg
Size of cereal seed (Wheat)	20,000 kg
Smaller than cereal seeds (Tomato)	10,000 kg
Large flower seeds (*Lathyrus*)	10,000 kg
Large tree seed (Quercus)	5,000 kg
Small flower seeds (*Achillea* spp.)	5,000 kg
Small tree seed (Eucalyptus)	1,000 kg

A fundamental problem in seed sampling is that it is almost impossible to obtain a perfectly uniform seed lot. A seed lot may also be packed in various ways. A seed lot may be a part of a shipload, it may be the entire content of a container, it may be packed in two or more larger containers, in many individual bags or several small containers. The lack of perfect uniformity, different packing and variable seed size makes it necessary to adopt different sampling methodologies and different sampling equipment in order to obtain a representative sample from any seed lot.

Once sampling is started, a specified number of primary samples are drawn from each seed lot. A primary sample is a small quantity of seed taken from a single place in the lot. At least the minimum number of primary samples taken from different parts of the lot are mixed to give a single large sample, the composite sample. The primary samples and composite sample are produced at the location of the seed consignment.

After thorough mixing of the composite sample, a submitted sample is send to the laboratory for testing. The quantity of a submitted sample must be at least equal to the minimum weight prescribed for the range of tests to be performed. The forwarded quantity is specified by ISTA and varies with the plant species. Examples are given in Table 2.

Once the submitted sample reaches the laboratory, working samples are produced by reduction of the submitted sample into smaller quantities of seed. One working sample is the amount of seed on which one laboratory test can be performed. The concept of a submitted sample may not be applied to plant quarantine as the personnel may make the composite sample themselves and bring it to the laboratory where working samples are made directly from the composite sample.

Table 2. Examples of size of submitted samples as prescribed by the International Seed Testing Association (ISTA, 1985)

Plant species	Minimum size of submitted sample
Quercus	500 seeds
Achillea spp.	5 g
Tomato	15 g
Eucalyptus	15-60 g
Lathyrus	400-600 g
Maize	1000 g
Wheat	1000 g

Drawing of primary samples

Primary samples may be drawn from containers, bags, small containers, from a bulk of seed or from a stream of seed. Equipment which is recommended and rules are described below.

Sampling in bags and small containers

The Nobbe trier, dynamic spear. The Nobbe trier or dynamic spear consist of a pointed tube, long enough to reach the centre of the container, with an oval hole near the pointed end. The trier is illustrated in Figure 1.

The trier is most convenient for sampling in bags. It is inserted into and reaching the centre of the bag at an angle to 30°. The hole at the tip of the trier should be facing downwards. The inserted trier is turned 180° along the longitudinal axis to make the hole facing up. The trier must be withdrawn immediately with a forward and backward vibratory motion and at decreasing speed. Seed is collected from the open end of the trier.

The size of the Nobbe trier depends on the seed size. A large trier can be used for seed of about the size of *Pisum sativum*. A standard trier is suitable for most chaffy grasses and seed about the size of *Hordeum vulgare*. A small Nobbe trier is suitable for small seeded grasses and other small seeds as *Brassica* spp. (Figure 1).

Sleeve trier or stick sampler. This trier may be used for sampling in larger containers as well as in bags. It consists of two metal tubes. The one fits loosely inside the other tube and is fitted with a handle. Both tubes have slots in their wall. When the slots are super-imposed by turning the inner tube, openings are formed which allow seed to flow into the cavity of the trier. The openings may lead into separate compartments of the tube or into a cavity of the length of the entire trier. The trier with compartments can be used in a vertical as well as horizontal position while the one-cavity trier can only be used in horizontal position.

Figur 1. Nobbe trier or dynamic spear with recommended specifications (from ISTA, 1986)

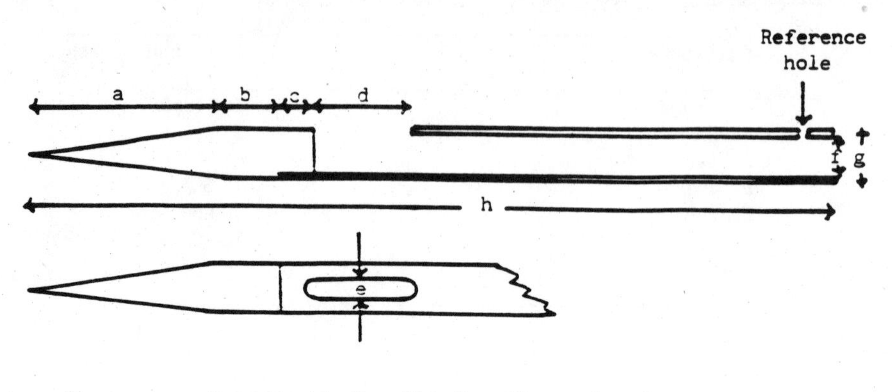

Size in mm	Point Shoulder Boss Hole Bore Diameter Length							
	(a)	(b)	(c)	(d)	(e)	(f)	(g)	(h)
Large trier	82	12	13	40	15	17	19	653
Standard trier	85	12	10	33	10.5	13	15	600
Small trier	42.5	7	8	20	7.5	20	12	432

A small trier has been found suitable for sampling small free-flowing seeds like *Trifolium* spp. and a larger trier has been found suitable for sampling of seed of the size of *Hordeum vulgare* and many other chaffy seeds (Figure 2).

In bags the trier with compartments is inserted diagonally from the top to the bottom with closed openings. Once in position inside the bag the compartments are opened by turning the inner tube. The compartments are filled completely while agitating the trier. Once filled, the openings are closed gently to avoid damage to the seed.

Sampling by hand. Primary samples are taken by hand by removing handfuls of seed from random positions. As it is difficult to force the hand deep into the seed, it may be necessary to partially empty the bags or containers in order to obtain primary samples from the lower layers of seed.

Sampling by hand may be the only satisfactory method for sampling under special circumstances and for sampling of very chaffy seed.

Sampling frequency. To ensure complete coverage of the seed lot, minimum sampling frequencies are prescribed by ISTA (Table 3). Of special interest to plant quarantine personnel is sampling of seed contained in small, closed tins or packets ready for retail distribution or small packets of germplasm.

Figure 2. The sleeve or stick trier with specifications (from ISTA, 1986)

Handle

Double steel tube with nine separate compartments

Seed size	length	outside diameter	slots	
	mm		mm	number
Small	762		12.7	9
Big	762		25.4	6

In the former case ISTA recommends combination of containers each weighing less than 50 kg into units of 100 kg. Each unit is then considered one container and the sampling frequency in Table 3 can be applied. In case of small sealed retail packets, each packet or tin in each unit of 100 kg is considered one primary sample.

In case of germplasm, each packet, irrespective of the size, is considered a seed lot and drawing of primary samples will usually not apply to such material. If the seed of the plant species in question is included in the legislation, each seed may be examined by the available non destructive seed health testing methods, inspected for required precautionary measures like seed treatment or imported based on the information in the phytosanitary certificate.

Sampling seed in loose bulk

By loose bulk means seed in high volume storage or transport containers or in heaps. If the depth of the seed does not exceed 2 metres, the triers described above may be used. The length of the triers should be long enough so that the bottom or centre of the seed volume can be reached. Drawing of seed lots exceeding 2 metre depth require special tools for sampling.

Neate sampler. This sampling device consist of a head in which the sample is contained and a steel tube which can be extended by tubes screwed together. The head has 260 mm back from the closed tip a hole measuring 25 x 52 mm. When pushing the sampler vertically into the seed, the hole is covered by a loose cover which due to friction keeps the hole closed but open the hole when the sampler is withdrawn. The collecting chamber fills rapidly. The collecting chamber is emptied through the extension tube. The minimum number of primary samples is prescribed by ISTA and is shown in Table 4.

Cargo sampler. This sampling tool is basically the same as the Neate sampler. The cargo sampler is different as the path to the collecting chamber is closed with a spring loaded valve. The minimum number of primary samples is the same as for the Neate sampler (Table 4).

Table 3. Sampling frequency for bags and other small containers (ISTA, 1986)

Number of containers in the seed lot	Minimum number of containers to be sampled
1 - 5 inclusive	Each container, from at least 5 positions
6 -14	Not less than 5 containers
15-30	At least one container in 3
31-49	Not less than 10 containers
50-400	At least one container in 5
401-560	Not less than 80 containers
561 or more	At least one container in 7

Table 4. Sampling frequency for seed loose in bulk (ISTA, 1986)

Size of the seed lot	Number of positions to be sampled
Up to 50 kg	Not less than 3
51 to 1500 kg	Not less than 5
1501 to 3000 kg	At least 1 for each 300 kg
3001 to 5000 kg	Not less than 10
5001 to 20000 kg	At least 1 for each 500 kg
20001 to 28000 kg	Not less than 40
28001 to 40000 kg	At least 1 for each 700 kg

Sampling from the seed stream

Sampling of large quantities of seed is often convenient when the seed is emptied out of the container. Primary samples are removed from the seed stream. The entire cross-section of the seed stream must be sampled uniformly each time a sample is drawn. Sampling may be done by hand or by various automatic devices.

Pelican sampler. The pelican sampler consists of a narrow wooden frame on which is mounted a bag. Each primary sample is drawn by passing the sampler through the stream of seed. The sampling frequency is shown in Table 4.

Preparation of working samples

After the required number of primary samples have been drawn the composite sample has to be prepared by thorough mixing of the primary samples. Particularly drawing of primary samples by automatic devices may result in a very large composite sample which cannot be mixed on laboratory machines. In that case it may be necessary to do repeated sampling of the composite sample by use of a Nobbe trier or sleeve type trier to make a second, reduced composite sample.

In plant quarantine there is usually no system of submitting seed samples to central laboratories. The seed is tested in the plant quarantine laboratory itself. Therefore the term 'submitted sample' does not apply in plant quarantine. Working samples are produced directly from the composite sample.

For mixing the composite sample and obtaining working samples the same equipment is used. Some manual procedures for sub-sampling are not suitable for mixing. The machines and manual sub-sampling procedures are described below.

Conical divider

The conical divider has no moving parts. A quantity of seed passed through the conical divider is divided into two halves. The seed is poured onto a hopper at the top of the divider. The bottom of the hopper is opened by a valve. By gravity, the seed pass over a cone below the hopper, pass into funnels and ultimately into two sample containers. Each alternate funnel end above one of the two containers. Mixing of a seed sample is carried out by passing the entire amount of seed two to three times through the divider. Working samples are produced by passing the seed through the divider and each time the seed in one sample container is removed until the required size of the working sample is reached.

Soil divider

The working principle of the soil divider is the same as for the conical divider. The seed is poured directly into the funnels which are arranged in a row instead of in a circle below a cone. Mixing and sub-sampling is the same as described for the conical divider.

Centrifugal divider

The centrifugal divider divides the seed sample into two equal sized parts by use of centrifugal force to distribute the seed through two outlet funnels to two sample containers. The machine has to be in an absolute horizontal position to work satisfactory. Mixing of a sample and sub-sampling is carried out as described for the dividers above.

Modified halving

By this technique the seed sample is divided into two halves with the use of a grid of cubical cells which fits onto a tray. Each alternate cell is closed at the bottom. The sample is halved by pouring the seed evenly over the grid with a side to side swing alternately in one direction and in a direction 90° to it. Half of the sample is trapped in the closed cells and half of the sample remains in the tray when removing the grid. The process is repeated until the working sample size is obtained.

For seed smaller than *Triticum* spp. the grid should consist of 144 cells each 25 x 25 x 25 mm^3.

Random cups

This method follow the same procedure as for modified halving. Instead of a grid the seed is poured over 8 cups placed at random in a tray. The working sample consists of the combined amount of seed collected in 6-8 cups. This technique is useful for production of working samples of less than 10 g of small seeds.

Spoon method

Preparation of working samples by the spoon method requires a clean tray of sufficient size, a spatula and a spoon with straight edge. The seed is poured into the tray as described for the modified halving method. By using the spoon and the spatula together, small quantities of seed is removed from at least five places selected at random in the tray. The combined spoonfuls is a working sample.

The spoon method is a very useful method in plant quarantine where other material than seed is also handled. The equipment is extremely simple. Any reasonably sized sample can be handled by selecting the proper size of tray and spoon. Careful pouring of the seed into the tray can reduce the release of dust including pesticide dust which is of particular significance when sampling treated seed. A simple suction device can remove the dust which is released even when pouring carefully.

Hand sampling

This method was removed in early 1950s from the list of methods recommended for obtaining working samples in the seed certification schemes as well as sampling of other items. The reason was that personal skills seemed to affect the sampling. However, the method was reintroduced in the ISTA rules in 1985 but only for a few selected genera with chaffy seeds.

The seed is poured onto a clean, smooth surface. The seed is divided into half by the aid of e.g. a ruler. Each half is divided repeatedly until two rows of each four equally sized heaps of seed are produced. Heap 1 and 3 in the first row are combined with heap 2 and 4 in the second row. The remaining four portions are removed. The process is repeated until a working sample is obtained.

References

ISTA. 1985. International rules for seed testing. *Seed Science and Technology* **13** (2): 300-520.
ISTA. 1986. *Handbook on Seed Sampling*. 61 pp.

Testing for Seed-borne Fungi

HENRIK JØRSKOV HANSEN and S.B. MATHUR

Danish Government Institute of Seed Pathology for Developing Countries, Ryvangs Allé 78, DK-2900 Hellerup, Denmark

Introduction

The International Seed Testing Association (ISTA, 1985) has made recommendations for detecting a few pathogens in a few hosts but apart from these, the testing procedure is the choice of individuals or institutions. In the following a few general testing procedures will be presented by which the majority of seed-borne fungi can be detected. In principle the testing procedures used in plant quarantine do not differ from the methods used in domestic seed certification programmes. However, the number of seeds tested in each test may be more in plant quarantine work in order to detect even trace infections. Furthermore, if more test methods are available for detection of a particular pathogen the most sensitive and/or selective procedure must be used in the plant quarantine work.

Testing methods for detection of seed-borne fungi in seed can be categorized as

- examination of seeds without incubation including examination of seed washings and extracted tissues
- examination of seeds after incubation
- examination of seedlings and plants and
- other methods

Examination of seeds without incubation

Inspection of dry seed

Inspection of dry seed is a seed health testing method which can be applied to detect fungal pathogens which when present in the seed may cause characteristic discoloration of the seed coat or changes in the seed size and shape. The method is also applicable for detection of fungi producing various visible structures in or on the seed or fungal structures mixed with the seed. In addition, inspection of dry seed also gives immediate information about insect and mechanical damage to the seed as well as whether the seeds have been treated with pesticides so that the samples are handled with appropriate precautions.

The inspection of dry seed in seed health testing is a qualitative test for which no standard working sample size has been worked out. However, it may be suggested to use a sample equal in size to the sample employed in the purity analysis in seed certification (from 0.5 g of *Achillea millefolium* to 1000 g of *Zea mays*).

All parts of a seed sample are examined carefully by naked eye and non seed matters are separated and identified. During the examination, emphasis is laid on galls, sclerotia and smut balls. By naked eye or under low magnification of a stereoscopic microscope the surface of the seeds are examined carefully for the presence of discolorations and fungal structures, including spores adhering to the surface or spore deposits.

Spores and other fungal structures are identified, characteristic discolorations are attributed to causal fungus while unspecific discolorations are noted.

Infection of seed may take place in a way that seed symptoms are not developed to the extent that they are visible at the time of harvest or fungal structures on the seed surface may be too small or too scattered to be detected directly. On the other hand, if symptoms or fungal structures are present, the causal fungus may be dead. Therefore, absence of symptoms or fungal structures in or with the seed does not generally imply freedom from infection while presence of symptoms also does not always imply presence of viable inoculum. Furthermore, a pathogen may be present in seed in one or more forms of which only one or few may be visible in the inspection of the dry seed.

Washing test

Washing test is a seed health testing method which is used to test seeds for those seed-borne pathogens of which the inoculum is present loosely on the seed surface.

The method is thus valid in particular for fungi causing smuts and bunts in graminaceous crops with the important exception of loose smut of wheat and barley. Among other fungi for which the method has been used are *Peronospora manshurica* (downy mildew of soybean) and *Protomyces macrosporus* (tumor disease of coriander).

The washing test is a qualitative test for which no standard working sample has been approved so far by ISTA. Various empirically fixed sample sizes have been proposed for different crop species, but none has so far been generally accepted.

The proceedure includes washing of a specified quantity of seed in water which may be added a few drops of detergent. Seed and water is shaken in order to release spores loosely attached to the seed surface as well as spores located in sori or similar structures covered by the seed coat tissue. The water is centrifuged and the sediment is examined on slides under the compound microscope. If a quantitative test is required, the analyst has to use one of the methods to count the number of spores present in a sample.

The washing test is a quick procedure for detection of surfaceborne spores of specified fungal pathogens. However, identification in this test is solely based on spore characteristics. The problems which can be faced is illustrated by the example of *Tilletia controversa* and *T. caries*.

Seed from gene banks may have been stored for years before they are dispatched, and much of the surfaceborne inoculum may be dead. However, many fungal spores, particularly amongst the smuts and bunts, are known to be able to survive for long time even under normal herbarium conditions (Neergaard, 1979). When stored in gene banks under the conditions prevailing in medium to long range storage, survival may be extended many fold. For such material, it may be necessary to test the viability of the detected spores. This may be done by use of the tetrazolium test as in the case of *Peronospora manshurica* (Pathak *et al.*, 1978).

Embryo count method

In a few diseases the mycelium of fungi which is present in the embryo has been found to be the only effective inoculum for producing disease in the field. Examples of this type are loose smut of wheat and barley (*Ustilago tritici* and *U. nuda*, respectively).

The fastest method for detecting their presence in the embryo is to extract the embryos and examine them under a microscope and identify them based on mycelial characteristics.

The embryo count method involves overnight soaking of seeds, usually 1000 in three replicates in an alkali (usually 5% sodium hydroxide) or an acid (concentrated sulphuric acid), separation of embryos from endosperm by washing with warm water (about 50°C), dehydration, separation from chaff, clearing by boiling in lactic acid and glycerol (1:2) and finally examination of embryos under stereomicroscope and checking of mycelium under a compound microscope. For details of the embryo count method, consult Shetty *et. al.* (1978), Rennie and Seaton (1975) and Khanzada *et. al.* (1980). The use of phenol can now be avoided in the embryo count method (Khanzada *et. al.*, submitted for publication 1992).

Examination of seeds after incubation

Blotter method

The blotter method is one of the incubation methods where seeds are plated on well water-soaked filter papers, incubated usually for 7 days at 20-25°C under 12 h alternating cycles of light [near ultraviolet (NUV) or fluorescent daylight] and darkness. Each seed, at the end of the incubation period, is examined thoroughly under different magnifications of a stereomicroscope (6-50 or 60 times) for the growth of fungi. Fungi found associated with seeds are carefully examined and identified based on "habit characters". Habit characters of fungi can only be learnt with practice.

In general, a working sample size used for the blotter method is 400 seeds. The number may be higher e.g. *Phoma lingam* in crucifer seeds (Maguire and Gabrielson, 1983)

The blotter method is widely used for detecting fungi which are able to produce mycelial growth and fruiting structures under the incubation conditions available in the test. The method is good in testing seeds for fungi such as *Alternaria*, *Ascochyta*, *Bipolaris*, *Botryodiplodia*, *Botrytis*, *Cercospora*, *Cladosporium*, *Colletotrichum*, *Curvularia*, *Drechslera*, *Fusarium*, *Macrophomina*, *Myrothecium*, *Phoma*, *Phomopsis*, *Rhizoctonia*, *Sclerotinia*. All kinds of seeds, cereals, vegetables, legumes, ornamentals and forest seeds, are tested by this method. It is the most widely practiced seed health testing method. Many laboratories use it as the first screening test for monitoring health condition of a seed lot.

Sometimes, the lids of petri dishes are found open at the end of the incubation period. This is due to pressure exerted by vigorously growing young seedlings. As a result of this opening, the filter papers become dry. Drying of filter papers has adverse effect on the development of fungi. Drying of filter papers can be avoided by putting a rubber band around the dish prior to incubation. The problem can also be avoided by killing the germs as first shown by Limonard (1966) in the Netherlands. The cereal seeds are sown on wet blotters, incubated for 24 hours at 20°C to induce germination. The germinating seeds are then frozen overnight at -20°C and subsequently incubated at 20°C in NUV for another five to seven days. The method is popularly called as deep-freeze blotter method.

The dead seedlings provide a good substrate for profuse development of fungi, e.g. *Fusarium* and *Septoria*.

Agar plate method

In the agar plate method, seed-borne fungi are detected and identified based on characteristics of colonies developing directly around infected seed on the agar.

Seeds are plated on various types of agar media including selective ones. The seed may or may not be surface sterilized with surface sterilizing agents before plating. After plating the seeds are incubated under conditions which favor growth and development of characteristic structures of the fungus for which the seed is tested.

The agar plate method can be used for almost all those seed-borne pathogens which are detectable in the blotter method. However, because the very different ability of different fungi to compete on artificial media only the fungi capable of growing under the given conditions will be recorded by this method.

The agar plate method provides an efficient tool for quick identification of specific seed infections once the analyst has become familiar with the colony characteristics of the fungi in question. Colony characteristics are likely to change if another agar medium is employed or if the incubation conditions are changed.

Considering the price of agar and time spend in preparing the medium the agar plate method may only be used in routine seed health testing for detection of fungi which are not easily recorded in the blotter test.

The standard working sample size for the agar plate method is 400 seeds.

Examination of seedlings and plants

Seedling symptom test

Some of the seed-borne fungi are capable of producing symptoms on young seedlings either in root, shoot or both. In some cases, the seedlings may even die from the effect of seed-borne inoculum. Some of the common examples of such fungal pathogens causing seedling diseases are species of *Alternaria*, *Ascochyta*, *Bipolaris*, *Fusarium*, *Drechslera*, *Colletotrichum*, *Macrophomina*, *Pyricularia*. Recording of symptoms in the seedlings are taken as a measure of infection present in the seed.

The seed can be sown in autoclaved soil, gravel, sand or similar material. Seed may also be sown on water agar. Infected seeds, under optimal conditions of temperature, light and humidity may develop symptoms comparable with those produced under field conditions.

The test is extremely beneficial for plant quarantine laboratories, particularly in handling small seed samples of germplasm. Seedlings with no symptoms should be transplanted in sterilized soil and grown to maturity under post-entry quarantine facilities. Any plant with symptoms of any kind must be discarded by burning, and seeds collected from healthy plants released, only. In case plants do not produce seeds under closed containers, healthy plants should be grown in post-entry quarantine isolation areas, inspected regularly before releasing seeds.

Growing-on test

This method is more close to field conditions. Greenhouse or growth-chamber facilities are required. Plants are grown in sterile media (sand, soil, vermiculite, etc). Environmental conditions should be favorable for the pathogens to be detected.

A major disadvantage of this method is the time required for growth of the host and expression of the symptoms which may in some cases take several weeks or even months. As a consequence, this method may be considered only when dealing with a pathogen on which very strict quarantine regulation has to be observed. An advantage is the possibility of saving seeds of valuable germplasm. Like in the seedling symptom test, the pathogens recorded in this test are seed transmitted whereas the other tests give information on presence or absence of seed-borne pathogens. The most important criterion in plant quarantine is to make sure that the pathogen is absent from the seed. In the growing-on test, however, if the environmental conditions are not favorable to disease development, the pathogen will not induce symptom development but it may come out in the substrate. If pots containing such symptom less plants are released to the importers it is likely that the substrate present in the pot can contaminate the area where such plants are grown. Therefore, the substrate in which the seed has been sown may contain inoculum of the pathogen, which however, did not transmit disease due to environmental factors. Therefore, plants raised in a seedling symptom test followed by the growing on test should not be released without additional precautions or only seed produced in the plant quarantine should be released.

References

ISTA. 1985. International rules for seed testing 1985. *Seed Science and Technology* **13**. No. 2.

Khanzada, A.K., Rennie, W.J., Mathur, S.B. and Neergaard, P. 1980. Evaluation of two routine embryo test procedures for assessing the incidence of loose smut infection in seed samples of wheat (*Triticum aestivum*). *Seed Science and Technology* **8**: 363-370.

Khanzada, A.K., Shetty, H.S., Mathur, S.B., Cappelli, C., Infantino, A. and Porta-Puglia, A. Avoidance of phenol in the embryo count procedure for detection of loose smut mycelium in wheat and barley. In press *Seed Science and Technology.*

Limonard, T. 1966. A modified blotter test for seed health. *Neth. J. Pl. Path.* **72**: 319-321.

Maguire, J.D. and Gabrielson, R.L. 1983. Testing techniques for *Phoma lingam. Seed Science and Technology* **11**: 599-605.

Neergaard, P. 1979. *Seed Pathology.* Vol. I and II. The Macmillan Press, London, UK.

Pathak, V.K., Mathur, S.B. and Neergaard, P. 1978. Detection of *Peronospora manshurica* (Naum.) Syd. in seeds of soybean, *Glycine max. EPPO Bulletin* **8**: 21-28.

Rennie, W.J. and Seaton, R.D. 1975: Loose smut of barley - the embryo test as a means of assessing loose smut infection in seed stocks. *Seed Science and Technology* **3**: 697-709.

Shetty, H.S., Khanzada, A.K., Mathur, S.B. and Neergaard, P. 1978. Procedures for detecting seed-borne inoculum of *Sclerospora graminicola* in pearl millet (*Pennisetum typhoides*). *Seed Science and Technology* **6**: 935-941.

Seed Health Testing for Bacteria*

H.K. MANANDHAR* and CARMEN N. MORTENSEN

*Central Division of Plant Pathology, Nepal Agricultural Research Council, Khumaltar, Lalitpur, Nepal;
**Danish Government Institute of Seed Pathology for Developing Countries, Ryvangs Allé 78, DK-2900 Hellerup, Denmark

Introduction

Seed health testing is required for determining the planting value of seeds, seed treatments to control the seed-borne infection, issuance of health certificates for quarantine. Such testing, especially for bacteria should effectively detect the pathogens since they are usually present at very low levels in seeds with big epidemic potentials. At the same time the test methods should be simple, reproducible, rapid and less-expensive.

Detection of bacteria in seed involves following three steps:

- extraction
- isolation and
- identification

Extraction of bacteria

Bacteria are extracted from seed by several methods. The selection of the method depends on where the bacteria are present, on the surface of seed or in the seed. It is also of importance to consider other microflora associated with the seed which may affect isolation of the target bacterium.

The three methods described here are:

1. soaking seeds or seed flour in liquid media (seed extractions)
2. plating seeds directly onto a agar medium (seed culturing) and
3. growing plants from seeds (plant culturing).

---

* This paper replaces Dr J. Taylor's paper presented at the Workshop.

Seed extractions

Bacteria are washed from seeds or extracted from seed flour by soaking briefly or for several hours in liquid media (saline, sterile water, buffers, enrichment medium etc.). Extractions can be made from surface-disinfested or from untreated seeds. Seed flour is obtained by grinding seeds dry in a blender. Incubation can takes place with or without shaking at low temperature or at room temperature. The seed extract can be used directly or after isolation on agar media for identification of bacteria.

Seed culturing

Seeds are directly plated onto a suitable, usually semi-selective agar medium plates. The thus plated seeds are incubated at about 25-28°C under darkness for 24 to 48 hours. Pseudomonads grow well at about 25°C while xanthomonads need higher temperature (ca 28°C). Bacterial colonies growing around infected seeds are isolated for identification.

Plant culturing

Plant culturing includes two methods, seedling symptom tests and growing-on tests. With the first method, seeds are sown on moist blotters, vermiculite, soil, sand or water agar. Seedlings are grown in greenhouses, growth chambers or a room with regulated conditions (light, humidity, temperature). Different plant parts including cotyledons are checked for symptom development and isolation is made from tissues showing symptoms.

In growing-on tests, plant are grown in the field or in greenhouse upto maturity. The plants are routinely inspected for the appearance of symptoms. Isolation is made from plant parts showing symptoms. This method is extremely important for testing small-sized seed samples, especially valuable germplasm by which seed can be harvested from healthy plants for release.

Isolation of bacteria

Isolation of bacteria involves two steps: 1) isolation of total bacterial population from seed extracts, seed cultures and plant parts showing symptoms, and 2) isolation of suspected colonies from the total bacterial population for identification. The first isolation (step 1) is usually made on a semi-selective medium while general medium is used for second isolation.

With some methods, e.g. immunofluorescence staining, steps for isolation can be skipped by using seed or plant extracts directly for identification. However, to obtain bacterial cultures for future use isolation is essential.

Identification of bacteria

Bacteria can be identified by different methods such as host plant inoculations, biochemical tests, serological tests, phage-plaque technique and or combination of them. Other methods like DNA probes, fatty acid profiles, dot-immunobinding (DIB) and others have also been used in the identification of bacteria, but not so common and practical for routine seed health testing, especially in developing countries. Various testing methods developed for detecting seed-borne pathogenic bacteria are presented in Table 2.

Host plant inoculations

Host plants, including ungerminated and germinating seeds are inoculated with seed extracts or bacterial suspensions in various ways e.g., plant injections, prickings, clippings, sprays, seed infiltrations etc. The inoculated plants or plant parts are incubated in chambers or in greenhouse under high relative humidity. Plants are checked for characteristic symptom development.

Difficulties may be encountered since similar symptoms are produced by different pathogenic bacteria, especially pseudomonads. If the conditions are not favourable, symptom expression may be checked or even false symptoms may be expressed. Therefore one must be sure and able to recognize characteristic symptoms of different host-pathogen combinations.

Biochemical tests

Biochemical tests such as Gram stain or KOH solubility, pigmentation, O/F metabolism, arginine dihydrolase, levan formation, Kovacs oxidase, gelatin hydrolysis, starch hydrolysis, lipase activity, nitrate reduction, utilization of carbon compounds, carbohydrates and organic acids (Lelliott and Stead, 1987; Schaad, 1988; Saettler *et al.,* 1989; Klement *et al.,* 1990) including growing bacteria on semi-selective media are common procedures for the identification of bacteria.

Use of semi-selective media allows a bacteria to grow with characteristic colony type which provides presumptive diagnosis of the bacteria. If necessary, it can be confirmed by other biochemical tests. Some semi-selective media useful for isolation of bacteria from seed are presented in Table 1.

Serological tests

Serology is perhaps the most useful tool for rapid and accurate diagnosis of seed-borne bacteria. The tests include aggultination (slide or tube), agar diffusion, immunofluorescence staining (direct and indirect IF), enzyme-linked immunosorbent assay (ELISA), dot-immunobinding (DIB) and others. Immunofluorescence assay in combination with agar plating are successfully used for the assay of many bacteria.

Table 1. Semi-selective media useful for isolation of bacteria from seed

Bacterium	Medium	References
Clavibacter michiganensis		
subsp. *michiganensis*	SCM, mSCM	Fatmi and Schaad, 1988, Waters and Bolkan, 1992
Curtobacterium flaccumfaciens		
pv. *flaccumfaciens*	NBY	Calzolari *et al.,* 1987
Erwinia stewartii	Nigrosin medium	Yifen *et al.,* 1981.
Pseudomonas solanacearum	SPA, TZC	Hayward, 1964; Kelman, 1954
Pseudomonas syringae		
pv. *glycinea*	M 71	Leben, 1972
pv. *phaseolicola*	MSP	Mohan and Schaad, 1987
pv. *pisi*	KBBC	Schaad and Mortensen, 1992
pv. *syringae*	KBBC	Mohan and Schaad, 1987
pv. *tomato*	KBBC, ND, YDC	Schaad and Mortensen, 1992; Bashan and Assouline, 1983
Xanthomonas campestris		
pv. *campestris*	NSCA, NSCAA, SX BSCAA, SM, FS	Schaad and Kendrick, 1975; Schaad and White, 1974; Randhawa and Schaad, 1984; Chun and Alvarez, 1983 *et al.,* 1974
pv. *carotae*	MD 5, XCS	Kuan *et al.,* 1985; Williford and Schaad, 1985
pv. *phaseoli*	MXP	Claflin *et al.,* 1987
pv. *translucens*	XTS	Schaad and Forster, 1985
pv. *vesicatoria*	Tween medium B, CKTM	McGuire *et al.,*1986; Sijam *et al.,* 1991
Xanthomonas oryzae		
pv. *oryzae*	XOS	Di *et al.,* 1991
pv. *oryzicola*	XOS	Di *et al.,* 1991

Phage-plaque technique

Phages (also known as bacteriophages) are virus particles which have the capacity of lysing bacterial cells. Phages of plant pathogenic bacteria are simple and rapid tools in bacterial identification, but sensitivity of the tests is greatly reduced when a large number of bacteria other than the target ones are present. This together with the lack of specificity and resistance of some strains may limit their application in routine testing. A phage culture is added to seed extracts and the liquid is tested. The phage titer is determined by formation of plaques on agar media or by the loss of turbidity of broth cultures.

Discussion

Several methods including highly specialized techniques (e.g. fatty acid profiles, DNA probes and others) have been developed for the detection of bacteria in seed. The choice of methods can be made according to the purpose of detection and availability of facilities and supplies. One of the most common practice is first to isolate the pathogen on semi-selective agar media and then to confirm its identity by one or different methods. Furthermore, the virulence of the isolates has to be proven. At present, semi-selective media and serology, especially indirect immunofluorescence, are extensively applied (Scortichini, 1991).

Although different seed assays for bacteria have been developed and used, the results from laboratory tests have seldom been correlated with disease incidence in the field (Schaad, 1984). There are few bacterial pathogens for which inoculum thresholds have been established, such as *Xanthomonas campestris* pv. *campestris* (Schaad *et al.,* 1980), *X. c.* pv. *translucens* (Schaad and Forster, 1993), *Pseudomonas syringae* pv. *phaseolicola* (Taylor *et al.,* 1979).

It is often said that unless results of laboratory seed assays correlate well with disease development in field, results of assays are questionable value (Schaad, 1984). But it depends on the purpose of the detection. For those bacterial pathogens which require zero tolerance (quarantine objects) the test methods must be highly sensitive so that a trace infection can be detected. In such cases there is no need of justifying correlation between results of seed assays and disease development in field.

Table 2. Assay methods used for detecting seed-borne pathogenic bacteria

Assay method(s)	Bacteria	Host	Reference
Growing-on and seedling symptom tests	*Pseudomonas avenae*	rice	Shakya and Chung, 1983
	Pseudomonas syringae		
	pv. *glycinea*	soybean	Parashar and Leben, 1972
	pv. *lachrymans*	cucumber	Volcani, 1966
	pv. *phaseolicola*	bean	Grogan and Kimble, 1967

....cont.

Assay method(s)	Bacteria	Host	Reference
.....cont.			
	pv. *pisi*	pea	Walker and Patel, 1964
	Xanthomonas campestris		
	pv. *campestris*	crucifers	Shackelton, 1962; Srinivasan *et al.*, 1973
	pv. *carotae*	carrot	Ark and Gardner, 1944
	pv. *malvacearum*	cotton	Halfon Meiri and Volacani,
1977			
	pv. *phaseoli*	bean	Shuster and Coyne, 1975
	pv. *vesicatoria*	pepper	Shekhawat and Chakravarti, 1979
	Xanthomonas oryzae		
	pv. *oryzae*	rice	Singh and Rao, 1977
Plant inoculation	*Clavibacter michiganensis*		
	subsp. *michiganensis*	tomato	Neergaard, 1979; Van Vaerenbergh and Chauveau, 1985
	Curtobacterium flaccumfaciens		
	pv. *flaccumfaciens*	bean	Calzolari *et al.*, 1987
	Pseudomonas syringae		
	pv. *glycinea*	soybean	Neergaard, 1979
	pv. *phaseolicola*	bean	Neergaard, 1979
	Xanthomonas campestris		
	pv. *campestris*	crucifers	Schaad, 1989
	pv. *phaseoli*	bean	Neergaard, 1979; Sheppard *et al.*, 1989
Seed inoculation	*Pseudomonas syringae*		
	pv. *glycinea*	soybean	Parashar and Leben, 1972
Seed culturing	*Pseudomonas fuscovaginae*	rice	Agarwal *et al.*, 1989
	Xanthomonas campestris		
	pv. *campestris*	crucifers	Schaad and Kendrick, 1975
	pv. *nigromaculans*	zinnia	Strider, 1979
	Xanthomonas oryzae		
	pv. *oryzae*	rice	Supriaman and Tantera, 1972
Seed extracts	*Clavibacter michiganensis*		
	subsp. *michiganensis*	tomato	Fatmi and Schaad, 1989
	Pseudomonas glumae	rice	Agarwal *et al.*, 1989
	Pseudomonas syringae		
	pv. *phaseolicola*	bean	Vuurde and Bovenkamp, 1987
	pv. *syringae*	bean	Mohan and Schaad, 1987
	pv. *tomato*	tomato	Jones *et al.*, 1989

.....cont

Assay method(s)	Bacteria	Host	Reference
.....cont.			
	Xanthomonas campestris		
	pv. *campestris*	crucifers	Lundsgaard, 1976; Schaad, 1989; Schaad and Donaldson, 1980
	pv. *carotae*	carrot	Kuan *et al.,* 1985; Kuan, 1989
	pv. *phaseoli*	bean	Ednie and Needham, 1973; Trujillo and Saettler, 1979; Sheppard *et al.,* 1989
	pv. *translucens*	wheat	Schaad and Forster, 1989
	pv. *vesicatoria*	tomato	McGuire and Jones, 1989
	Xanthomonas oryzae		
	pv. *oryzae*	rice	Di *et al.,* 1991
	pv. *oryzicola*	rice	Di *et al.,* 1991
Phage-plaque technique	*Clavibacter michiganensis*		
	subsp. *michiganensis*	tomato	Neergaard, 1979
	Pseudomonas syringae		
	pv. *atrofaciens*	cereals	Neergaard, 1979
	pv. *phaseolicola*	bean	Neergaard, 1979
	pv. *pisi*	pea	Neergaard, 1979
	Xanthomonas campestris		
	pv. *phaseoli*	bean	Katznelson, 1950; Sheppard *et al.,* 1989
	Xanthomonas oryzae		
	pv. *oryzae*	rice	Wakimoto, 1954
Serology	*Erwinia stewartii*	corn	Lamka *et al.,* 1991
	Pseudomonas syringae		
	pv. *phaseolicola*	bean	Vuurde and Bovenkamp, 1989
	Pseudomonas syringae		
	pv. *tomato*	tomato	Jones *et al.,* 1989
	Xanthomonas campestris		
	pv. *campestris*	crucifers	Schaad, 1978 and 1989
	pv. *phaseoli*	bean	Sheppard *et al.,* 1989

References

Agarwal, P.C., Mortensen, C.N., and Mathur, S.B. 1989. *Seed-borne Diseases and Seed Health Testing of Rice*. Danish Government Institute of Seed Pathology for Developing Countries, Technical Bulletin No. 3 and CAB International Mycological Institute, Phytopathological Papers 30. 106 pp.

Ark, P. A. and Gardner, M.W. 1944. Carrot blight as it affects the roots. *Phytopathology* **34:** 416-420.

Bashan, Y. and Assouline, I. 1983. Complementary bacterial enrichment techniques for the detection of *Pseudomonas syringae* pv. *tomato* and *Xanthomonas campestris* pv. *vesicatoria* in infested tomato and pepper seeds. *Phytoparasitica* **11:** 187-193.

Calzolari, A., Tomesani, M. and Mazzucchi. 1987. Comparision of immunofluorescence staining and indirect isolation for the detection of *Corynebacterium flaccumfaciens* in bean seeds. *Bulletin OEPP/EPPO* **17:** 157-163.

Chun, W.W.C. and Alvarez, A.M. 1983. A starch-methionine medium for isolation of *Xanthomonas campestris* pv. *campestris* from plant debris in soil. *Plant Disease* **67:** 632-635.

Clafin, L.E., Vidaver, A.K. and Sasser, M. 1987. MXP, a semi-selective medium for *Xanthomonas campestris* pv. *phaseoli. Phytopathology* **77:** 730-734.

Coleno, A. 1968. Utilization de technique d'immunofluorescence pour le depistage de *Pseudomonas phaseolicola* (Burk) Dowson dans les lots de semences de harocot contaminés. *Comptes rendus Séances de l' Academie d' Agriculture de France* **54:** 1016-1020.

Di, M., Ye, H., Schaad, N.W. and Roth, D.A. 1991. Selective recovery of *Xanthomonas* spp. from rice seed. *Phytopathology* **81:** 1358-1363.

Ednie, A.B. and Needham, S.M. 1973. Laboratory test for internally-borne *Xanthomonas phaseoli* and *Xanthomonas phaseoli* var. *fuscans* in the field bean (*Phaseolus vulgaris*L.). In: *Proceedings of the Association of Official Seed Analysts of North America* **63:** 76-82.

Fatmi, M. and Schaad, N.W. 1988. Semi-selective medium for isolation of *Clavibacter michiganense* subsp. *michiganense* from tomato seed. *Phytopathology* **78:** 121-126.

Fatmi, M. and Schaad, N.W. 1989. Detection of *Clavibacter michiganense* subsp. *michiganense* in tomato seed. pp. 45-49. In: *Detection of Bacteria in Seed* (eds. A.W. Saettler, N.W. Schaad and D.A. Roth). APS Press. The American Phytopathological Society, St. Paul, Minnesota.

Franken, A.A.J.M. and Van den Bovenkamp, G.W. 1989. The application of the combined use of immunofluorescence microscopy and dilution-plating to detect *Pseudomonas syringae* pv. *pisi* in pea seeds. pp. 871-876. In: *Plant Pathogenic Bacteria* (ed. Z. Klement). Part B. Proceedings of the 7th International Conference on Plant Pathogenic Bacteria. Budapest, Hungary, June 11-16, 1989. Akadémiae Kiadó, Budapest.

Grogan, R.G. and Kimble, K.A. 1967. The role of seed contamination in the transmission of *Pseudomonas phaseolicola* in *Phaseolus vulgaris. Phytopathology* **57:** 28-31.

Guthrie, J.W., Dean, L.L., Butcher, C.L., Fenwick, H.S. and Finley, A.M. 1975. The epidemiology and control of halo blight in Idaho. *Agric. Exp. St., Universaity of Idaho Bull* **550**. 11 pp.

Halfon-Meiri, A. and Volani, Z. 1977. A combined method for detecting *Colletotrichum* and *Xanthomonas malvacearum* in cotton seed. *Seed Science and Technology* **5:** 129-139.

Hayward, A.C. 1964. Characteristics of *Pseudomonas solanacearum. J. Appl. Bacteriology* **27:** 265-2

Jones, J.B., McCarter, S.M. and Gitaitis, R.D. 1989. Detection of *Pseudomonas syringae* pv. *tomato* in tomato. pp. 50-58. In: *Detection of Bacteria in Seed* (eds. A.W. Saettler, N.W. Schaad and D.A. Roth). APS Press. The American Phytopathological Society, St. Paul, Minnesota.

Katznelson, H. 1950. The detection of internally-borne bacterial pathogens of beans by a rapid phage-plaque count technique. *Science* **112:** 645-647.

Kelman, A. 1954. The relationship of pathogenicity in *Pseudomonas solanacearum* to colony appearance on a tetrazolium medium. *Phytopathology* **44:** 693-695.

Klement, Z., Rudolph, K. and Sands, D.C. (eds.). 1990. *Methods in Phytobacteriology*. Akadémiai Kiadó, Budapest.

Kuan, T.L. 1989. Detection of *Xanthomonas campestris* pv. *carotae* in carrot. pp. 63-67. In: *Detection of Bacteria in Seed.* APS Press. The American Phytopathological Society, St. Paul, Minnesota.

Kuan, T.L., Minsavage, G.V., and Gabrielson, R.L. 1985. Detection of *Xanthomonas campestris* pv. *carotae* in carrot seed. *Plant Disease* **69:** 758-760.

Lamka, G.L., Hill, J.H., McGee, D.C. and Braun, E.J. 1991. Development of an immunosorbent assay for seedborne *Erwinia stewartii* in corn seeds. *Phytopathology* **81:** 839-846.

Leben, C. 1972. The development of a selective medium for *Pseudomonas glycinea. Phytopathology* **62:** 674-676.

Lelliott, R.A. and Stead, D.E. 1987. *Methods for the Diagnosis of Bacterial Diseases of Plants. Methods in Plant Pathology* (ed. T.F. Preece). Vol. 2. Blackwell Scientific Publications. 216 pp.

Lundsgaard, T. 1976. A method for detection of *Xanthomonas campestris* (Pammel) Dowson in Brassica seeds. *Statens Plantetilsyn* **21:** 34-38.

McGuire, R.G. and Jones, J.B. 1989. Detection of *Xanthomonas campestris* pv. *vesicatoria* in tomato. pp. 59-62. In: *Detection of Bacteria in Seed* (eds. A.W. Saettler, N.W. Schaad and D.A. Roth). APS Press. The American Phytopathological Society, St. Paul, Minnesota.

Maguire, R.G., Jones, J.B. and Saaaer, M. 1986. Tween media for the semiselective isolation of *Xanthomonas campestris* pv. *vesicatoria* from soil and plant material. *Plant Disease* **70:** 887-891.

Mohan, S.K. and Schaad, N.W. 1987. An improved agar plating assay for detecting *Pseudomonas syringae* pv. *syringae* and *Pseudomonas syringae* pv. *phaseolicola* in contaminated bean seed. *Phytopathology* **77:** 1390-1395.

Neergaard, P. 1979. *Seed Pathology*. Vol. I and II. MacMillan Press, London.

Parashar, R.D. and Leben, C. 1972. Detection of *Pseudomonas glycinea* in soybean seed lots. *Phytopathology* **62:** 1075-1077.

Randhawa, P.S. and Schaad, N.W. 1984. Selective isolation of *Xanthomonas campestris* pv. *campestris* from crucifer seeds. *Phytopathology* **74:** 268-272.

Saettler, A.W., Schaad, N.W., and Roth, D.A. (eds.). 1989. *Detection of Bacteria in Seed.* APS Press. 122 pp.

Schaad, N.W. 1978. Use of direct and indirect immunofluorescence tests for identification of *Xanthomonas campestris. Phytopathology* **68:** 249-252.

Schaad, N.W. 1989. Detection of *Xanthomonas campestris* pv. *campestris* in crucifers. pp. 68-75. In: *Detection of Bacteria in Seed* (eds. A.W. Saettler, N.W. Schaad and D.A. Roth). APS Press. The American Phytopathological Society, St. Paul, Minnesota.

Schaad, N.W. (ed.). 1988. *Laboratory Guide for Identification of Plant Pathogenic Bacteria.* 2nd edition. APS Press. The American Phytopathological Society, St. Paul, Minnesota. 164 pp.

Schaad, N.W. and Donaldson, R.C. 1980. Comparision of two methods for detection of *Xanthomonas campestris* in infected crucifer seeds. *Seed Science and Technology* **8:** 383-392.

Schaad, N.W. and Forster, R.L. 1985. A semiselective agar medium for isolating *Xanthomonas campestris* pv. *translucens* from wheat seeds. *Phytopathology* **75:** 260-263.

Schaad, N.W. and Forster, R.L. 1989. Detection of *Xanthomonas campestris* pv. *translucens* in wheat. pp. 41-44. In: *Detection of Bacteria in Seed* (eds. A.W. Saettler, N.W. Schaad and D.A. Roth). APS Press. The American Phytopathological Society, St. Paul, Minnesota.

Schaad, N.W. and Forster, R.L. 1993. Black chaff (*Xanthomonas campestris* pv. *translucens*). pp. 129-136. In: *Seed-borne Diseases and Seed Health Testing of Wheat* (eds. S.B. Mathur and B.M. Cunfer). Danish Government Institute of Seed Pathology for Developing Countries, Denmark.

Schaad, N.W. and Kendrick, R. 1975. A qualitative method of detecting *Xanthomonas campestris* in crucifer seed. *Phytopathology* **65:** 1034-1036.

Schaad, N.W. and Mortensen, C.N. 1992. Detection of seed-borne bacteria. pp. 199-206. In: *Seed Pathology* (eds. S.B. Mathur and J. Jørgensen). Proceedings of the CTA Seminar held at Copenhagen, Denmark on 20-25 June 1988. Technical Centre for Agricultural and Rural Cooperation, CTA and Danish Government Institute of Seed Pathology for Developing Countries.

Schaad, N.W. and White, W.C. 1974. A selective medium for soil isolation and enumeration of *Xanthomonas campestris. Phytopathology* **64:** 876-880.

Schuster, M.L. and Coyne, D.P. 1975. Detection of bacteria in bean seed, *Xanthomonas phaseoli. Ann. Rep. Bean. Imp. Coop.* **18:** 71.

Scortichini, M. 1991. Diagnosis of seed-borne bacterial diseases. *Petria* **1**, *Supplemento* **1:** 31-45.

Shackelton, D.A. 1962. A method of detection of *Xanthomonas campestris* (Pammel 1895) Dowson, 1939, in Brassica seed. *Nature* **193:** 78.

Shakya, D.D. and Chung, H.S. 1983. Detection of *Pseudomonas avenae* in rice. *Seed Science and Technology* **11:** 583-588.

Shekhawat, P.S. and Chakravarti, B.P. 1979. Comparision of agar plate and cotyledon methods for detection of *Xanthomonas vesicatoria* in chilli seeds. *Phytopathology* Z **94:** 80-84.

Sheppard, J.W., Roth, D.A. and Saettler, A.W. 1989. Detection of *Xanthomonas campestris* pv. *phaseoli* in bean. pp. 17-29. In: *Detection of Bacteria in Seed* (eds. A.W. Saettler, N.W. Schaad and D.A. Roth). APS Press. The American Phytopathological Society, St. Paul, Minnesota.

Sijam, K., Chang, C.J. and Gitaitis, R.D. 1991. An agar medium for the isolation and identification of *Xanthomonas campestris* pv. *vesicatoria* from seed. *Phytopathology* **81:** 831-834.

Singh, R.A. and Rao, M.H.S. 1977. A simple technique for detecting *Xanthomonas oryzae* in rice seeds. *Seed Science and Technology* **5:** 123-127.

Srinivasan, M.C., Neergaard, P. and Mathur, S.B. 1973. A technique for detection of *Xanthomonas campestris* in routine seed health testing of crucifers. *Seed Science and Technology* **1:** 853-859.

Strider, D.L. 1979. Detection of *Xanthomonas nigromaculans* f.sp. *zinniae* in zinnia seed. *Plant Disease Reporter* **63:** 869-873.

Supriaman, J. and Tantera, D.M. 1972. Detection of *Xanthomonas oryzae* in rice seed samples. *Contr. Cent. Res. Inst. Agric. Bogor No.* **1.** 14 pp.

Trujillo, G.E. and Saettler, A.W. 1979. A combined semi-selective medium and serology test for the detection of *Xanthomonas* blight bacteria in bean seed. *J. Seed Technology* **4:** 35-41.

Van Vaerenbergh, J. and Chauveau, J.F. 1985. Host plant inoculations for the detection of (latent) *Corynebacterium michiganense* (E.F. Smith) Jensen. *Med. Fac. Landbouww. Rijksuniv. Gent* **50/3a:** 973-995.

Volcani, Z. 1966. A quantitative method for assessing cucumber seed infection caused by *Pseudomonas lachrymans*. *Israel Journal of Botany* **15:** 192-197.

Vuurde, J.W.L. van and Bovenkamp, G.W. van den. 1987. *Pseudomonas syringae* pv. *phaseolicola*. *ISTA Handbook on Seed Health Testing Working Sheet No.* **65**. International Seed testing Association, Zurich. 8 pp.

Wakimoto, S. 1954. The determination of the presence of *Xanthomonas oryzae* by phage technique. *Sci. Bull. Fac. Agric. Kyushu Univ.* **14:** 495-498.

Walker, J.C. and Patel, P.N. 1964. Inheritance of resistance to halo blight infection and disease transmission in halo blight of bean. *Phytopathology* **54:** 952-954.

Waters, C.M. and Bolkan, H.A. 1992. An improved semi-selective medium and method of extraction for detecting *Clavibacter michiganensis* subsp. *michiganensis* in tomato seeds. *Phytopathology* **82:** 1072.

Yifen, G., Zaiqun, L. and Dafu, Yu. 1981. A selective medium for isolation of *Erwinia stewartii* (E.F. Smith) Dye from imported corn. *Kexue Tongbao* **26:** 630 [in Chinese].

Detection of Seed-borne Viruses in Seeds

K.M. MAKKOUK* and L. BOS**

*International Center for Agricultural Research in the Dry Areas (ICARDA), P.O.Box 5466, Aleppo, Syria; **Research Institute for Plant Protection (IPO-DLO), P.O.Box 9060, NL-6700 GW Wageningen, The Netherlands.

Introduction

Prerequisite to seed testing for virus infection is knowledge about the relationship between the virus to be detected and the seed tissues, as well as about the mechanism of seed transmission (Bos and Makkouk, this volume; Bos, 1989). Seed-coat abnormalities and the presence of virus and viral products (e.g. coat protein) only indicate mother plant infection and the chance of later seedling infection with a few highly infectious viruses (tobamoviruses, southern bean mosaic virus). Several viruses may not be able to infect the embryo and naturally reach the seedling, and they may not survive seed-coat maturation. Remnant viral coat proteins, nevertheless, lead to "false positive" reaction in highly sensitive serological tests when whole seeds are tested.

For general information on virus transmission via seed and on seed testing for virus infection see Bos (1977), Mandahar (1981) and, Agarwal and Sinclair (1987).

Testing techniques for virus detection should be simple, sensitive, reliable and non-expensive. It might not be feasible that a single test has all the above features, but the progress made in diagnostics over the last decade made it possible for some tests, such as ELISA, to meet most of the above criteria. In this communication only the tests which are widely used for virus detection in seeds are discussed.

Methods of seed testing

Visual examination

Seeds originating from plants infected with some seed-borne viruses (e.g. broad bean stain virus) may have symptoms suggestive of virus infection. However, virus symptoms on seed coats only indicate that such seeds originated from infected plants, but do not prove that the virus has reached the embryo. In addition, seeds containing virus may look completely normal. Accordingly, examination of seed coats for virus symptoms has little value, if any, with respect to actual seed transmission.

Growing-on test

The growing out of seeds to allow visual examination of seedlings for virus symptoms is a very common test used in many places, sometimes in combination with other tests. The test is carried out by planting a number of seeds (100-400) in soil. Symptoms produced are recorded 2-3 weeks after emergence. Such test is best performed in an insect-proof greenhouse where temperature could be controlled around 25°C. Light is another important factor and light intensity of around 10,000 foot candles produces the best results. Higher temperatures and lower light intensities will produce poor viral symptoms which makes the test less reliable. However, in some cases the viral symptoms may be completely masked in spite of the optimal growth conditions, and many seed-borne viruses are symptomless in several of their hosts. Accordingly the growing-on tests are often used in combination with an infectivity or other direct test.

Infectivity test

The indicators used in this test should be highly susceptible and of sensitive cultivars. Seeds are ground with 0.01 M (or 0.03 M) phosphate buffer, pH 7.0. Sodium nitrate (0.5%) and activated charcoal (75 mg/ml) are added and the suspension is directly inoculated onto the leaves previously dusted with carborundum. Tests plants are kept under observation for some weeks depending on the virus(es) expected. If indicator plants which produce local lesions were used, results may be obtained in 2-3 days. Seeds can thus be tested individually, but in routine tests they are usually tested in groups of e.g. 10. Some viruses are still detected in single infected seeds ground together with 500 or more uninfected seeds (as of lettuce mosaic virus in lettuce seeds) but this has to be investigated first for each virus/host combination. The same test could be carried out on seedlings developing from seeds to be tested.

Serological tests

Serological tests are useful for large-scale testing. They have reached high sensitivity and efficiency which permits testing seeds in groups of 10, 100 or sometimes more. Seed coats may have to be removed before serological tests to avoid false positives. There are several serological tests which could be employed for seed testing. In this communication we will emphasize three most commonly used methods, namely agglutination test, enzyme-linked immunosorbent assay (ELISA) and immunospecific electron microscopy (ISEM).

a. Agglutination tests

There are two types of agglutination tests which could be employed.

Latex tests. In this test antibodies are adsorbed onto commercially available polystyrene latex particles. Optimal attachment of antibodies to the latex occurs only within a narrow range of concentration of antiserum or purified immunoglobulins. Using antibody-coated latex, several workers succeeded in detecting 100 to 1000 fold smaller quantities of virus than was possible by the gel double diffusion (Van Regenmortel, 1982).

A suspension of latex particles (about 0.8 µm diameter, Difco Laboratories) is diluted 1:15 with saline, and equal volumes of this suspension and suitably diluted purified immunoglobulin are incubated for 1 hour. The optimal dilution of the globulin fraction of antiserum to be used for sensitization lies in the region 1:50 to 1:2000 and is determined by trial and error. The diluent is 0.05 M Tris-HCl buffer, pH 7.2. The sensitized latex is recovered by low- speed centrifugation and washed twice in an equal volume of 0.05 M Tris-HCl buffer, pH 7.2, containing 0.02% polyvinylpyrrolidone as a stabilizer. The final sediment is suspended in the same buffer containing 0.02% sodium azide, and is brought to a volume identical to that of the diluted immunoglobulin originally used. The sensitized latex preparation can be stored for more than 6 months in the cold without any loss in activity.

Virobacterial agglutination test. The virobacterial agglutination test for the identification of plant viruses was first described by Chirkov *et al.*, (1984) and used more recently by Walkey *et al.* (1989). The technique is simple and allows rapid identification of viruses from very small volumes of infected crude sap.

A suspension of formalin-treated *Staphylococcus aureus* (1 vol) is mixed with diluted virus antiserum (1 vol 1:1 antiserum/glycerol mixture + 24 vol phosphate buffered saline, pH 7.2 containing 2 mg/ml sodium azide) (5 vol). This bacterial - virus antibody conjugate is coloured by the addition of several drops of saturated alcoholic basic fuchsin. Approximately 4 µl of this conjugate is mixed with 2 µl of the crude sap to be tested on a Welled multitest slide (Flow Laboratories). A suitable negative control using healthy host sap must also be prepared. A positive reaction is indicated by agglutination of the bacterial particles within 1-3 minutes.

b. Enzyme-linked immunosorbent assay (ELISA)

In this method (Clark and Adams, 1977), the plate wells are first coated with antivirus gamma globulin (IgG) and the virus in the test sample is then trapped by the adsorbed antibody. The presence of virus is revealed by an enzyme-labeled anti-virus conjugate. The various stages of the test are performed in polystyrene microtiter plates. If bacterial contamination is likely to occur 0.02% sodium azide may be added to all the buffers used in the assay. The wells are coated at 37°C by incubation (15 min to several hours) with 1-10 µg/ml antivirus globulins (300 µl) diluted in 0.05 M sodium carbonate buffer, pH 9.6. The globulins can be prepared by precipitation from antiserum with an equal volume of 4 M ammonium sulfate. There appears to be no advantage of using highly purified IgG. Plates are then rinsed three times with phosphate-buffered saline pH 7.4, containing 0.05% Tween-20 (PBS-T), and once with 1% bovine serum albumin in PBS-T. This last washing step may not be necessary in all systems.

Antigen preparations (300 µl) in PBS-T are added to the coated wells for 1-16 hours at 37°C. The addition to crude plant extracts of 1-2% polyvinyl pyrolidone, 1 M urea, and reducing agents may reduce nonspecific reactions and increase the sensitivity of virus detection. With unstable viruses that are easily degraded in PBS-T, it may be advantageous to combine the antigen incubation step with the subsequent incubation with conjugate. The virus extract is first mixed with the diluted conjugate, before being added to the wells.

After rinsing the wells, the antivirus enzyme conjugate diluted in PBS-T (300 µl) is added for 1-3 hours at 37°C. The enzyme conjugate most commonly used is prepared with alkaline phospholactase (Boehringer, Mannheim, or Sigma St. Louis, Missouri) by coupling the globulins with enzyme at ratios ranging between 1:1.5 and 8:1 (w/w globulin:enzyme) using 0.06% glutaraldehyde. The optimal dilution of conjugate (usually 1:200-1:1000) must be determined empirically. The conjugate should be stored at 4°C in the presence of 1% bovine serum albumin.

After further rinsing, the bound enzyme conjugate is detected by the addition of 200 µl of the substrate p-nitrophenyl phosphate at 1 mg/ml in 0.1 M diethanolamine buffer, pH 9.8. After 1-3 hours of hydrolysis, the reaction is stopped by the addition of 50 µl of 3 M NaOH to each well. Results are scored visually by the appearance of a yellow colour, or absorbance at 405 nm are read in a spectrophotometer. Results are considered positive if the absorbance is twice that found with healthy controls or alternatively, if it is 3 standard deviation units higher than the mean of a negative control.

Since the first report on the use of ELISA for the detection of plant viruses in 1977 a number of ELISA variants to increase test efficiency were reported. More recently, Van den Heuvel and Peters (1989) reported an improved procedure where ELISA sensitivity and rapidity can be significantly improved. Using this techniques, 50-100 pg were reported to be detected in a sample. Such a procedure could prove very instrumental in detecting viruses which are present in very low concentration in seeds or seedlings.

Another recent modification is the use of penicillinase (PNC)-based ELISA (Sudarshana and Reddy, 1989). The sensitivity of this test was found to be comparable with the alkaline phosphatase system. The main advantage of the PNC-based ELISA is that the substrate, penicillin, is readily available in developing countries, and at a much lower cost than p-nitrophenyl phosphate.

c. Immunospecific electron microscopy (ISEM)

This method combines the high resolution of electron microscopy and the specificity of serological reaction. The method could be carried out by either coating virus particles with antibodies, commonly referred to as decoration (Milne and Luisoni, 1977), or trapping virus particles on electron microscope grids already coated with antibodies (Derrick, 1973). ISEM is one of the most sensitive methods for virus detection, but it is expensive and not fit for routine use, especially in the developing countries.

Concluding remarks

All methods of seed testing have their own merits, and the final choice may be a matter of compromise between simplicity, rapidity, reliability, sensitivity, and cost. Serological testing is rapidly increasing because of its sensitivity and suitability for large-scale and routine application. But sometimes these tests are too sensitive, in that they may detect viruses or even non-infective viral coat protein in the seed coat while there is no virus in the embryo. That is why proper precautions against false positive reactions must be taken and why it is often useful to rely on two tests simultaneously such as serology in addition to growing-on test or infectivity assay.

Virus concentration in single seeds may vary. Therefore, samples should not be too large. Sometimes the seeds must first germinate and the young sprouts or seedlings be tested to increase virus concentration and detectability. Furthermore, the virus may not be evenly distributed in the embryo; it may be detectable in the sprout but not in the cotyledons, or vice-versa.

Thus far, there is no method to remove virus from seed embryos once these are infected. Most of the tests described here are destructive i.e. seeds that proved virus-free are lost in the procedure. In order to secure virus-free seed, e.g. during quarantine, for release, seed must be produced on plants free of infection. Small numbers can pass through quarantine. The seeds must then be sown in a vector-free glasshouse and the resulting seedlings be individually tested. From those seedlings that are found to be free of the infection, the seeds can be harvested and released for distribution and for breeding purposes. To select seed of groundnut free of peanut mottle and stripe viruses at ICRISAT a non-destructive test has been developed whereby thin slices are cut out from the cotyledonary part of individual seeds and tested serologically in groups of 25. If reaction was positive, the 25 seeds can be individually retested and groups or individual seeds free of the two viruses may be released for entry or export (Bharatan *et al.*, 1984; D.V.R. Deddy, personal communication, 1990). Of course, guarantees only cover viruses for which the material was tested. There may be other seed-transmitted viruses that are yet to be detected. Biological tests using a range of test species may be needed to reveal their presence.

In the case of international transfer of commercial seed, at points of entry, it is only possible to test samples representing the consignment. Virus freedom of samples of 2500 seeds only guarantee a rate of infection below 0.1 %. Therefore seed sample testing cannot guarantee absolute freedom of infection.

References

Agarwal, V.K. and Sinclair J.B. 1987. Testing methods for seed-borne viruses. pp. 56-67. In: *Principles of Seed Pathology*. Vol. II. CRC Press, Inc. Boca Raton, Florida.

Bharathan, N., Reddy D.V.R., Rajeshwari, R., Murthy, V.K., Rao, V.R. and Lister, R.M. 1984. Screening peanut germplasm lines by enzyme-linked immunosorbent assay for seed-transmission of peanut mottle virus. *Plant Disease* **68**: 757-758.

Bos, L. 1977. Seed-borne viruses. pp. 39-69. In: *Plant Health and Quarantine in International Transfer of Genetic Resources* (eds. W.B. Hewitt and L. Chiarappa). CRC Press. Inc.

Bos, L. 1989. Virus transmission via seed: mechanisms, detection and implications for quality and quarantine. pp. 115-118. In: *Introduction of Germplasm and Plant Quarantine Procedures.* (eds. A.H. Jalil *et al.*). ASEAN PLANTI, Malaysia.

Chirkov, S.N., Olovnikov, A.M., Surguchyova, N.A. and Atabekov, J.G. 1984. Immunodiagnosis of plant viruses by a virobacterial agglutination test. *Annals of applied Biology* **104**: 477-483.

Clark, M.F. and Adams, A.N. 1977. Characteristics of the microplate method of enzyme-linked immunosorbent assay for the detection of plant viruses. *Journal of General Virology* **34**: 475-483.

Derrick, K.S. 1973. Quantitative assay for plant viruses using serologically specific electron microscopy. *Virology* **56**: 652-653.

Mandahar, C.L. 1981. Virus transmission through seed and pollen. pp. 241-292. In: *Plant Diseases and Vectors: Ecology and Epidemiology.* (eds. K. Maramorosch and K.F. Harris). Academic Press, New York.

Milne, R.G. and Luisoni, E. 1977. Rapid immune electron microscopy of virus preparations. pp. 265-281. In: *Methods in Virology.* (eds. K. Maramorosch and H. Koprowski). Vol. VI. Academic Press, New York.

Sudarshana, M.R. and Reddy, D.V.R. 1990. Penicillinase-based enzyme-linked immunosorbent assay for the detection of plant viruses. *Journal of Virological Methods* **26**: 45-52.

Van den Heuvel, J.F.J.M. and Peters, D. 1989. Improved detection of potato leaf role virus in plant material and in aphids. *Phytopathology* **79**: 963-967.

Van Regenmortel, M.H.V. 1982. *Serology and Immunochemistry of Plant Viruses.* Academic Press, New York. 302 pp.

Walkey, D.G.A., Lyons, N.F. and Taylor, J.D. 1989. A rapid slide agglutination test for detecting plant viruses using *Staphylococcus aureus* virus antibody conjugation. Paper presented at the *Meeting of BSPP/BCPC/ University of East Anglia, England.* Abstract book **19**.

Seed Health Testing for Nematodes

GEORGES CAUBEL

Institute National Recherche Agronomique, Laboratoire de Zoologie, F 35650 Le Rheu, France

Introduction

The nematodes may be attached to the seed or they may be found under the seed coat, penetrating the cotyledons or mixed with plant debris. So selection of extraction methods depends on nematodes and plant species involved.

Direct examination

When symptoms are recognizable on seeds e.g. galls of *Anguina tritici* in wheat small amounts of seeds can be examined directly under a stereoscopic microscope using transmitted light. The material is examined under water in a Petri dish or watchglass and dissected with needles. Nematodes released from the tissues can be collected with a needle and transferred to slides for a more detailed observation.

In the case of *Vicia* bean, heavily infested seeds may have split seed coat and necrotic patches appear on the cotyledons. Sometimes, masses of nematodes form "nematode-wool" on the surface of the seed, especially in the slit at the hilum region. But seeds without symptoms may also contain numerous *D. dipsaci*.

Extraction methods

Baermann funnel technique (after Hooper, 1970)

The seeds are placed in muslin or on a supporting sieve which is immersed in water in a funnel, or other suitable container. This funnel, with a piece of rubber tubing attached to the stem and closed by a screw clip, is placed in a support stand. Seed samples, or debris, should be spread thinly on flat plastic sieves with nylon mesh. A filter paper can be used if the seed contains fine debris. It is often helpful to add to the water a trace of wetting agent and a bactericide solution. The nematodes emerge and pass through the cloth or filter paper and sink to the bottom of the funnel stem. Nematodes that have settled in the tubing are collected in a beaker. A disadvantage of the Baermann funnel technique is poor oxygenation, especially at the base of the funnel where the nematodes collect. Therefore it is preferable to collect the extracted nematodes as soon as possible. The number of nematodes recovered can be affected by breakdown products from seeds, bacterial activity, lack of oxygen, temperature and length of storage. Two days, in general, are required for adequate nematode recovery.

Soaking method

Place about 100 g of the seed sample in a washing bowl or a pail, with two litres of water. Soak the sample overnight. Then the suspension is shaken for a few seconds and poured onto a set of two sieves, 1000 and 10 mesh, respectively; the first holds back seeds and debris, which are washed and put into the washing bowl again. The residue on the 10-mesh sieve is quickly and carefully recovered in a beaker for observing under a stereoscopic microscope. The water is recovered in a second bowl and again poured onto the 10-micron sieve and put back into the first bowl. This procedure is repeated four times.

In the case of *Vicia* bean, extraction may be obtained by leaving the seeds in water for 10-25 hours. The contents are shaken and the water decanted into a beaker. The seeds are rinsed with 100 ml of water which is added to the beaker. The extract is allowed to settle for 4 hours before the water is carefully decanted. The sediment is poured into a dish for examination. The described extraction method will extract only rather superficially borne nematodes; not all individuals borne inside the cotyledons may be released. Soaked tissues release products, causing the water to become cloudy, and often contamination with fungi can be observed after a few days. If seed stocks contain a fungicide, observation of suspensions may be difficult.

Mist extraction system (after Seinhorst, 1950)

A continuous fine mist of water is sprayed over the infested material. Active nematodes emerge and can be recovered from the water which collects below. Nematodes recovered are often active because oxygenation is good. The method is suitable for recovering most motile nematodes. A proportion of some *Aphelenchoides* spp. which swim, may be lost with the overflowing water. A spray nozzle passing about 5 litres of water per hour under pressure is commonly used. Seeds are placed on a milk filter or paper tissue supported on a coarse sieve. Sometimes a finer sieve may be used without a filter. The sieve is placed inside a funnel, the stem of which leads to the bottom of a beaker of about 200 ml capacity. The rate of overflow must not be too high to carry a significant number of nematodes with it.

Extracted nematodes may be cleaned and inactive ones eliminated by allowing them to pass through a paper tissue. By this method active specimens are recovered.

The original method has been modified by Hooper (1970) in a "mist chamber"; the author of this paper proposed a "double funnel" method, where the water collects in a second funnel. Excess water flows away via a small overflow pipe. Nematodes concentrate in a small glass tube fitted to the funnel by a piece of rubber tubing.

The funnel-spray method combines the original Baermann funnel and the mist method of Seinhorst: one day is generally sufficient to obtain satisfactory results from small samples of about 50 g and the extraction of stem eelworms usually takes about five days.

Other methods

The centrifugal floatation method, often used to extract nematodes from root tissues, is seldom used for the extraction of nematodes from seeds.

In some cases, it would also be possible to detect nematodes in a biological test, for example after germination of seeds if early symptoms are developed. But it is only used to confirm an earlier test.

In the future, it should be possible to use other detection methods. For example, the X-ray radiography technique is a fast and commonly used method which allows the printing on film of the internal structure of an object or living matter. Its application to seeds is now well-known and it permits the control of seed quality to guide the lot conditioning straight after harvesting, checking of pelleted seeds, homogenization of seed samples, etc. It is also possible to detect infestation of seed of *Vicia* bean by stem nematode.

Identification of nematode species

Preparation and mounting

Microscopic observations are made on living and dead specimens in water or in special media. In order to kill, fix and prepare temporary and permanent slide mounts, we carry out the following operations:

1) Kill and fix the specimens with the formaldehyde-acetic acid (4:1) fixative

- prepare the fixative as follows:
  + formalin (ca. 40% formaldehyde) 10 ml
  + glacial acetic acid 1 ml
  + distilled or deionized water 89 ml
- place nematode in a drop of water
- add hot fixative to the drop and leave undisturbed for a few hours. Specimens can be left in this fixative.

2) Mount the nematodes
For temporary mounts, place 10-15 specimens in a drop of fixative on a clean glass microscope slide. Pick up the specimens with a bamboo splinter, a nylon bristle attached to a small handle. Seal the cover glass by applying nail polish

with a small brush. This is a quick means of examination. These slides will keep for several days or weeks.

For permanent mounts, glycerine is the best medium. It is the best way to keep a reference slide collection, but the process is rather long (Seinhorst, 1959). Such a slide should last for 20 years.

Identification

To identify genera, a magnification of 30 to 60 times is normally used. Higher magnification is required to distinguish phytoparasitic nematodes from occasional but very common non-phytophagous species which do not bear a mouth stylet. To facilitate examination of individuals, they may be transferred to a drop of water with a needle and covered by a cover slip before gentle heating over a flame. This makes the nematode straighten and immobile. Different morphological criteria are used: mensurations, head structure, buccal stylet, oesophagus and genital structures. This is a matter for a specialist (see CIH descriptions of plant-parasitic nematodes and ISTA sheets for description).

To identify stem races, we may use length of individuals. Two main types of races occur. Adults more than 1.6 mm long are of the so-called "giant race". A normal race measures from 1.0 to 1.5 mm and the giant race from 1.6 to 2.3 mm.

References

Franklin, M. 1972. *Aphelenchoides besseyi. CIH Description of Plant-parasitic Nematodes.* Set **1** No. 4. Commonwealth Institute of Parasitology. St. Albans, Herts., England.

Franklin, M. 1972. *Heterodera schachtii. CIH Description of Plant-parasitic Nematodes.* Set **1** No. 1 Commonwealth Institute of Parasitology. St. Albans, Herts., England.

Hooper, D.J. 1970. *Laboratory Methods for Work with Plant and Soil Nematodes.* (ed. J.F. Southey). Tech Bull. **2**. Ministry of Agriculture and Fish and Food, London.

Hooper, D.J. 1972. *Ditylenchus dipsaci. CIH Descriptions of Plant Parasitic Nematodes.* Set **1** No. 14. Commonwealth Institute of Parasitology. St. Albans, Herts., England.

Hooper, D.J. 1973. *Ditylenchus destructor. CIH Descriptions of Plant Parasitic Nematodes.* Set **2** No. 21. Commonwealth Institute of Parasitology. St. Albans, Herts., England.

Hooper, D.J. 1975. *Aphelenchoides blastophthorus. CIH Descriptions of Plant Parasitic Nematodes.* Set **5** No. 73. Commonwealth Institute of Parasitology. St. Albans, Herts., England.

ISTA Handbook on Seed Health Testing. Working sheets: No. 52, 54, and 57. Zürich, Switzerland.

Seinhorst, J.W. 1950. De betekenis van de toestand van de grond voor hept optreden van aanstating door het stengelaaltje. *Tijdschr. Plantenziekten* **56**: 289-348.

Seinhorst, J.W. 1959. *Nematologica* **4**: 67-69.

Seshadri, A.R. 1975. *Ditylenchus angustus. CIH Descriptions of Plant-parasitic Nematodes.* Set **5** No. 64. Commonwealth Institute of Parasitology. St. Albans, Herts., England.

Southey, J.F. 1973. *Anguina agrotis. CIH Descriptions of Plant-parasitic Nematodes.* Set **2** No. 20. Commonwealth Institute of Parasitology. St. Albans, Herts., England.

Southey, J.F. 1974. *Anguina graminis. CIH Descriptions of Plant-parasitic Nematodes.* Set **4** No. 53. Commonwealth Institute of Parasitology. St. Albans, Herts., England.

Southey, J.F. 1985. *Anguina tritici. CIH Descriptions of Plant-parasitic Nematodes.* Set **1** No. 13. Commonwealth Institute of Parasitology. St. Albans, Herts., England.

Seed Health Testing for Insects

KLAUS RICHTER

University of Leipzig, Institute of Tropical Agriculture, Fichtestr. 28, D-7030 Leipzig, Germany

Introduction

The ultimate success of any strategy for pest control and quarantine requirements depends on the effectiveness of the methods used for the detection of pests. Storage facilities, kind of seed and climatic factors are closely related to testing procedures. Moreover, the behaviour of different species in response to their environment also has a profound influence on the choice of the methods to use.

Monitoring and trapping methods

For testing insect pests numerous methods are known. An overview of the common methods in relation to the behavior of the insects is given in Table 1. For selection and modification of suitable methods, which are adapted to the specific product-pest combination it could be useful to reduce each method to their basic reaction (Richter, 1975).

Using the basic principle as it is shown in Table 2, it is possible:

- to select the most effective method for specific purposes
- to modify the prescription in relation to the actual situation and
- to combine different methods for the final assessment of the seed lot.

Monitoring

Early detection of stored pests can prevent the rapid spread of damage, and is of special importance for preserving the quality of seed during storage.

Therefore it is worth underlining that monitoring must be a constant programme to maintain the quality of stored seed.

Widely used method for detecting insect infestation in different storage facilities is the visual inspection. Easily noticeable by this method are moths and strong flying coleopterans, but also adults and larvae which hide in bag stacks. Moreover, secondary products as dust, smell etc. can be of great service for detecting the presence of insects.

Table 1. Methods for detection of insect pests in stored seed and storage facilities

Free living pests		Hidden infestation	
Physical methods	Chemical or biochemical methods	Physical methods	Chemical or biochemical methods
Sieving	Pheromone traps	Flotation	Use of stains
Flotation	Food traps	X-ray technique	CO_2 - detection
Tullgren funnels		CT - technique	Determination of amino acids
Pitfall traps		Sound detection	
Suction traps			
Light traps			

Although visual inspection gives information about insect infestation, it "cannot be standardized sufficiently to make it entirely reliable for comparison between different situations and different workers, because it is largely subjective" (Peng and Morallo-Rejesus, 1987).

Trapping methods

The progress in the development of trapping techniques have resulted in traps enhanced by pheromones or food attractants which are effective for the detection of a wide range of insect pests (Pinniger, 1990).

Various types of traps baited with pheromones or food attractants are available for important species of lepidopterous pests (e.g., *Phyctiidae*) as well as for *Trogoderma, Tribolium*, *Rhizopertha*, *Prostephanus*, *Sitophilus*, *Cryptolestes* and *Lasioderma*. According to Peng and Morallo-Rejesus (1987) the types of traps in use may be classified as follows:

Card traps. Strips of corrugated cardboard inserted between layers of bags provide a very useful and repeatable method of monitoring infestation changes. The strips provide refuge sites into which moth larvae (*Corcyra* spp. and *Ephestia* spp.) crawl before pupation. Many beetle species including *Tribolium*, *Oryzaephilus*, *Cryptolestes*, *Ahasverus*, and *Carpophilus* also aggregate in these traps.

Pitfall traps. These are vessels made from glass or plastics, open or covered by perforated lids and can be baited with food or pheromones for better assessment. They are inserted into the bulk grain, where the openings are at the same level as the grain surface. The trap is constructed in such a way that the caught insects cannot escape.

Sticky traps. Commercial "fly papers" or strips of plastic fastened to boards with a non-hardening gum make very good traps for flying insects and is probably one of the best early warning trapping systems available at low costs.

Light traps. They attract a wide range of insect species, and the wave length of the light used influences the trapping results.

Food traps. Small bags made of cloth or plastic mesh cloth, filled with a suitable mixture of food grains, can quickly detect insect infestation from the surroundings.

Suction traps They can collect a broad spectrum of insects but are relatively expensive.

Pheromone traps. Pheromones as attractive agents can be used in special pheromone traps or along with other insect traps (e.g., cardboard traps). Combinations of pheromone lures and food baits are also possible and highly effective for recording a variety of pests at very low population levels (Burkholder and Ma, 1985).

In relation to other testing procedures (sampling and examination by laboratory methods) Hillman (1990) found that pitfall traps baited with aggregation pheromone or without baits detected *Sitophilus, Tribolium*, and *Cryptolestes* earlier than did samples taken from a 1.8 m compartmentalized grain trier.

Although much research work must be done, the newest results in monitoring research, presented at the 5th International Working Conference on Stored-Product Protection held in 1990 at Bordeaux, show that the use of traps can be recommended for monitoring storage facilities.

Sampling and laboratory detection methods

When the condition in the store are not favourable for the movement of insects or the attractivity of any bait or pheromone is not high enough, trapping methods give insufficient or no results. In these cases or if the establishment of an infestation degree (number of pests per kg of the product) are necessary, the most reliable method is by examining the stored seed itself.

For the assessment of infestation of seed and for estimating the level of infestation the three steps required are: proper sampling, separation of developmental stages of insects that live outside the kernels, and detection of internal infestation.

Sampling

The samples are taken by means of different kinds of samplers. The bag samplers must reach to the centre of the bag and should have the below mentioned diameters:

- for small seed 12 mm
- for cereals 15 mm
- for grain legumes 20 mm

Grain samplers with or without intermediary sections and probes are longer than bag samplers and mostly used for taking samples from bulk grain.

The sampling procedure must ensure that the tested sample is representative of the entire bulk. The number of primary samples therefore depends solely on the number of bags, regardless of their weight. According to Gwinner *et al.* (1990) the minimum number of primary samples should be:

- for small stacks of bags

up to 10 bags	one sample from each bag
11 to 25 bags	one sample from every second bag
26 to 50 bags	one sample from every third bag

- for stacks with a large number of bags

51 to 100 bags	10 primary samples
300 to 400 bags	20 primary samples
800 to 900 bags	30 primary samples
1400 to 1600 bags	40 primary samples
2000 to 2500 bags	50 primary samples

The number of primary samples in stacks containing more than 10,000 bags can be calculated as the square root of the number of bags in the stack.

Separation of free living pests from samples

For testing seed samples with regard to live stages outside the kernels, sieving is the only recommendable method. The diameter of the sieve meshes depends on the size of the seeds (e.g., for wheat 2 to 3 mm). For practical purposes the following combination of sieves with different diameters of meshes should be used:

1 = upper sieve	(meshes wider than seed diameter)
2 = second sieve	(meshes smaller than seed diameter)
3 = third sieve	(not general in use; for separation of different groups of pests according to their size).
4 = drawer	

The sieving procedure has to include also the standardization of shaking time and frequency. Some free living stages of insect species, such as *Trogoderma* larvae or adults and larvae of the genus *Cryptolestes*, are difficult to sieve from grain, since they are clinging tightly to the kernels (Smith, 1977).

Detection of internal infestation

From Table 2 it is evident that there are numerous possibilities for detecting developmental stages of insects that live inside the kernels. For utilization of one or more of these methods attention must be paid to the following aspects:

- all stages of the life cycle from egg to the adult may exist within the grain, and must be detected,
- early detection must be ensured by the selected method,
- for practical purposes the procedures should be easy and rapidly applicable, except, special techniques, and
- a high reliability of the results obtained is strongly required.

To detect initial internal infestation the use of **methods for direct visualization** gives the most accurate results. Tiny areas of tissue with different internal structure are traceable by the modern CT-techniques based on the use of X-rays or nuclear magnetic resonance spectroscopy (Stein 1986; Chambers *et al.*,1984). The equipment constructed for human health purposes can be used for examination of plant material without problems (Richter, 1989).

Sound detection methods. These methods are early described by Adams *et al.* in 1954. The methods are sensitive and are capable of giving safe information about living stages. Their detection capacity reach approximately 1 insect (moving or feeding stage) per kg of the product (Anonymous, 1973). In a number of countries research work to improve this method is going on.

Flotation and staining methods. These methods are cheap and easy to apply. Newly infested seed is detectable by flotation a few days after oviposition with salt solutions which have specific gravity equivalent to the healthy product (Richter and Tchalale, 1991). The earliest possible detection of *Sitophilus* larvae was found 5 to 6 days after oviposition on triticale.

Increasing density of the flotation medium from 1 to 1.24 g/l tested at the pest-product combination wheat/*S. oryzae* has also led to an earlier detection. The stain commonly used is acid fuchsin whereby the gelatinous weevil egg plug of *Sitophilus* species in grain were stained a bright cherry red (Frankenfeld, 1948). Other punctures produced by boring larvae (e.g., 1st instar larvae from *R. dominica* and *S. cerealella* get a light pink color.

The holes bored by young bruchid larvae can be stained by a 1% solution of iodine followed by treatment with a 0.5 % solution of KOH and rinsing. The openings of the tunnels are stained in black (Bartos and Verner, 1990).

Standardization of the testing procedure

Taking into consideration the above mentioned points it will be evident that standardization of the methods is of great importance for the comparability of the results obtained under different conditions. Standardization leads to harmonization in the quarantine regulations between countries with similar or dissimilar agricultural systems.

Taking into account the experience collected in Germany, a recommendable combination of methods could be:

1. Sieving method for separation of the free living developmental stages from the seed grain
2. Flotation method using salt solutions
3. Staining method for the detection of egg plugs and feeding and webbing punctures

The third method is important, because the detectability of eggs and 1st instar larvae of the flotation technique is insufficient. The second and third methods are necessary when the testing procedure with the sieving method gives negative results.

Table 2. Classification of detection methods according to their basic principle

DIFFERENCE BETWEEN healthy product and pest or infested product		basic principle for detection	METHOD(S)
SIZE			
big	small	separation	sieving method
small	big		
DENSITY			
high	low	separation	cracking flotation method flotation method suction traps
COMPOSITION OF PROTEIN			
insect protein absent	insect protein present	determination	use of stains ninhydrin process Ashman-Simon Infestation Detector
INTERNAL STRUCTURE			
homogenous	nonhomogeneous	determination	different methods for direct visualization X-ray technique CT-technique
CARBON DIOXIDE OUTPUT			
low	high	measurement	chemical and physical methods for determination of CO_2
MOBILITY			
not mobile	mobile	separation	different trapping methods BT-funnels
MOBILITY			
not mobile	mobile	measurement	sound detection

References

Adams, R.E., Wolfe, J.E., Milner, M. and Shellenberger, J.A. 1954. Detection of internal insect infestation in grain by sound amplification. *Cereal Chem.* **31**: 271-276

Anonymous 1973. Insectofon - an instrument to detect insects in seed and grain. *ISTA News Bulletin* No. **42**.

Bartos, J. and Verner, H.P. 1990. *Vorratsschädlinge*. Berlin. 232 pp.

Burkholder, W.E. and Ma, M. 1985. Pheromones for monitoring and control of stored-product insects. *Annual Review of Entomology* **30**: 257-272

Chambers, J., McKevitt, N.J. and Stubbs, M.R. 1984. Nuclear magnetic resonance spectroscopy for the development and the detection of the grain weevil, *Sitophilus granarius* (L.) (Coleoptera:Curculionidae) within wheat kernels. *Bull. entom. Res.* **74**: 707-724.

Frankenfeld, J.C. 1948. Staining methods for detecting weevil infestation in grain. USDA Agricultural Research Administration, Bureau of Entomology and Plant Quarantine ET-256, 4 pp.

Gwinner, J., Harnisch. R. and Mück, O. 1990. *Manual on the Prevention of Post Harvest Losses*. Hamburg. 294 pp.

Hillmann, R.C. 1990. Performance and grower acceptance of pitfall traps used to monitor Coleoptera that infest farm-stored corn in northeastern North Carolina. Paper presented at the *5th International Working Conference on Stored-Product Protection*. Bordeaux.

Peng, W.K. and Morallo-Rejesus, B. 1987. Grain Storage Insects. pp. 163-178. In: *Proceedings of the International Workshop in Rice Seed Health*. Manila: 163-178.

Pinniger, D.B. 1990. Sampling and trapping insect populations, the importance of environment, insects and trade. Paper presented at the *5th International Working Conference on Stored- Product Protection*. Bordeaux.

Richter, K. 1975. *Vorratsschutz in der Pflanzenproduktion*. Leipzig.205 pp.

Richter, K. 1989. Nachweismethoden für Vorratsschädlinge. *Schriftenreihe des WB Pflanzen- und Vorratsschutz,Institut für tropische Landwirtschaft*. (in press).

Richter, K. und Tchalale, P. 1991. Zur Eignung der Flotationsmethode für die Früherkennung des Innenbefalls von Getreide mit *Sitophilus oryzae* (L.) (Coleoptera:Curculionidae). un-published.

Smith, L.B. (1977). Efficiency of Berlese-Tullgren funnels for removal of the Rusty Grain Beetle *Cryptolestes ferrugineus*, from wheat samples. *Canad. Entomol.* **109**: 503-509.

Stein, W. 1986. *Vorratsschädlinge und Hausungeziefer*. Stuttgart.

Monitoring of Treated Seed

HENRIK JØRSKOV HANSEN

Danish Government Institute of Seed Pathology for Developing Countries, Ryvangs Allé 78, DK-2900 Hellerup, Denmark

Introduction

The main purpose of plant quarantine is to act as a safety filter which allow plant material into a region without simultaneous import of pathogens which are considered a threat to economically important crops of the importing region or country.

The seed which pass through the plant quarantine organization falls in three main categories which are food grain or fodder, planting material and germplasm. Food grain and fodder will not be dealt with since other precautionary measures other than seed treatment will usually be taken if any required.

Seed imported as planting material in general is produced under controlled conditions and the place of production is known. A range of precautions are taken right from the time of production in order to fulfill the general quality requirements including high productivity as well as the quarantine requirements before import. A seed treatment is frequently the last step in the efforts to eradicate possible trace infection of the pathogen(s) considered of quarantine significance in the importing country as well as to reduce infection by other seed-borne quality diseases.

Seed imported as germplasm is used in plant breeding and high productivity is not essential. Germplasm is usually dispatched from a known place but the place and conditions of production may not be known and the risk of carrying exotic pests of quarantine significance is higher than for seed as planting material. Germplasm is frequently distributed from one institution to most of the world. In order to reduce the risk of import of disease causing organisms, the seed is treated. The seed treatment may be given in the exporting country, in the importing country or in third country quarantine.

Quality of seed treatment

Depending on the type of seed treatment it has an effect on both the seed tissues and the pest. Seed treatments are in general effective in controlling pests like fungi, nematodes and insects. Also bacteria may be affected when antibiotics are employed but virus in the seed may not be affected. The quality of the seed treatment is dependent on the relative effect on the seed and the pathogen. Basically the selection of seed treatment and mechanics involved in treating seed is the same whether for planting material or germplasm. However, the quantity but particularly the use and origin of each consignment of planting material including 'nurseries' for experimental purpose and germplasm is different and therefore the seed treatment quality concept of planting material differs from the quality concept in germplasm.

A high quality seed treatment of planting material reduce the seed-borne inoculum of quality diseases, eradicate the seed-borne organisms of quarantine significance and retain the high germination capability of the seed as well as the high productivity of the emerging plant.

A high quality seed treatment of germplasm should eradicate any seed-borne organism of quarantine significance, maybe even on the expense of high germination capability and high productivity. Therefore germplasm is often treated with a larger dose of chemical, even reaching some phytotoxic level, than seed for plant production.

Monitoring of treated seed

Seed treatments may be divided into two main groups. Those which leave a deposit on the seed surface and those which does not leave a deposit on the surface (Table 1).

The quality of seed treatments leaving no deposit on the surface of the seed can only be monitored by testing the direct effect on the seed-borne pathogen and the seed tissue by using ordinary testing procedures for seed health and seed quality.

The quality of seed treatments leaving visible or invisible deposits on the surface can be monitored as mentioned above as well as indirectly by testing the quality of the deposit left on the seed surface.

In this way monitoring of treated seed contains four important topics:

- testing for the effect of the treatment on
    - a) the host
    - b) the pathogen and
    - c) disease development
- detection of deposits on seed surfaces
- testing of the quality of the deposit and
- precautions by the analyst against exposure to the deposit.

Table 1. Methods of seed treatment

Method of application	Deposit on the seed surface
Steep treatment: soaking in solution or suspension	Little or none
Slurry treatment: dust plus liquid in soup-like slurry	Considerable amount
Quick-wet treatment: concentrated usually volatile liquid	Some.
Coating: dust suspended in adhesive material	Some to considerable
Dust treatment: seed mixed with dry dust	Considerable amount
Pelleting	Large amount
Fumigation: vapour	None
Hot water treatment: No chemicals involved	None

In this paper only detection of deposits on seed surfaces, testing of the quality of the deposit and precautions to be taken by the analyst against exposure to the deposit will be discussed. Testing of the quality of the deposit will be limited to deposits of fungicides.

The factors on which the quality of a seed treatment with fungicides is dependent are:

- selection of effective chemical
- overall correct dosage for the seed lot
- uniform dosage on each seed and
- even distribution on the seed surface

A range of methods are available to test one or more of the quality factors. As the tests has to be used in routine work, it is a must that the method is quick, easy and cheap to perform and that it gives reliable results.

Table 2. Examples of different test-organisms used in bio-assay for different pesticides

Microorganism	Pesticides	Vehicle	Reference
Arthrobacter globiformis	Maneb, mancozeb, zineb	Cereals	Jørgensen, J. (Pers. comm.)
A. strain AR18	Mercurial compounds, captan, thiram	wheat	Ehle, 1971
Aspergillus niger	Mercurial compounds, captan, thiram phaltan, dichlone	Cereals, barley, oats, wheat, peanut, rye	Halfon-Meiri & Dishon, 1964; Crosier et. al., 1961
Bacillus cerrus	Lindan, thiram, captan, bromophos	Cereals	Jørgensen, J. (Pers. comm.)
B. subtilis	Mercurial compounds captan, thiram, aldrin, lindane	paper discs, bean, cabbage, radish, pea, tomato	Pepper & Clausen, 1963
Cladosporium herbarum	Organomercurials	wheat	Blair, 1950
Clavibacter rathayi	Mercurial compounds, captan, thiram	barley, wheat, beet Barley, wheat, beet	Kovacs, 1963
Drechslera sorokiniana	Mercurial compounds, chloranil, thiram, imazilil, thiabendazol, prochloraz	Wheat	Mead, 1945; Jørgensen (pers. comm)
Fusarium nivale	Triadimenol + rabenzazol + fuberidazol	rye	Sethofer, 1946 - 1947; Jørgensen (pers. comm)
Fusarium spp.	Organomercurials	Cereals	
Glomerella cinguiata	Organomercurials Organomercurials, thiram Mercurial compounds Thiram Captan Hexachlorobenzene Phaltan, dichlone, pentachloronitrobenzene, difolatan	wheat, Radish, maize, pea, soybean, vetch seed	Blair, 1950; Samra, 1956; Arny, 1952; Crosier & Bruce, 1960, Crosier et. al. 1961; Leben & Keitt, 1950; Richardson, 1966
13 fungi isolated from groundnuts	Captan, thiram	Paper discs	
Myrothecium verrucaria	Mercurial compounds, thiram, captan, phaltan, dichlone	Rye	Crosier et. al., 1961
Penicillium atrovenetum	Benomyl, chloroneb, carboxin	Soybean seedling tissue	Thapliyal & Sinclair, 1971
P. digitatum	Organomercurials	Cereals	Jackson & Ballard, 1973
P. expansum	Benomyl	Soybean	Ellis & Sinclair, 1976
P. italicum	Systemics	Paper discs,	Edginton et. al., 1973
P. purpurogenum	Mercurial compounds	Wheat	Machecek, 1950
P. thomii	Mercurial compounds, dexon, thiram	Wheat	
Rhizoctonia solani	Organomercurials, captan, benomyl, non-mercurial mixtures, ethirimol Thiabendazole, TH 7462, carboxin, chloroneb	Cereals, cotton seedling homogenate, soybean seedling tissue	Allam et. al., 1969; Martin, 1975; Thapliyal & Sinclair, 1971
Saccharomyces pastorianus	Captan	pea seedlings, paper discs	Wallen & Hoffmann, 1959
Sarcina lutea	Mercurial compounds, captan, thiram, aldrin, dexon	Wheat, barley, corn, bean, wheat	Wallen & Hoffmann, 1959 Molinas, 1961; Smith & Crosier, 1966
Sporobolomyces roseus	Carbendazim, guanoctine	Cereals	Jørgensen (pers. comm)
Stemphylium ilicis	Organomercurials	sugar beet	Byford, 1975
Ustilago maydis	Systemics	Paper discs	Edgington et. al., 1973

Methods

Measurement of dye intensity

Most of all fungicides contain dye (Radtke, 1982), which is clearly seen on the surface of treated seed, to avoid fatal use of the treated seed as food or feed.

Direct visual examination of colour intensity on treated seeds is useful as a first crude evaluation of the amount and distribution of fungicide on the seeds.

It is possible to determine the average amount of fungicide supplied to the seed lot by extracting the dye and measure the dye content in the cleared extract by specto-photometric methods or, more crude, by visual comparison of colour intensity of the extract with a standard colour series.

Reliable results from indirect determination of supplied fungicide through dye-extraction is only obtained if the control institution is supplied with a part of the very formulation with which the seed has been treated (Radtke, 1982). The reason to do so is that the amount of dye may vary from one lot of production to another even by the same company. Radtke (1982) showed a difference in dye content up to 11 per cent. Sampling, duration of extraction, extraction medium and measuring equipment may cause variation of ±15 per cent. Based on this information and if the tolerance limit for correctly treated seed is fixed at ±10 per cent, measurements should be allowed to fluctuate up to ±25 per cent.

Extraction of dye followed by measurement of dye intensity may be a quick method which, however, only shows the overall dosage of fungicide in the seed sample representing a seed lot.

Paper disc method

With this method the active ingredient of the formulated chemical supplied to the seeds are extracted. Paper discs with a specified fluid holding capacity are dipped into the extract and placed on agar inoculated with a specified test organism. After incubation for a specific period the size of the inhibition zone formed around each disc indicates the relative amount of active ingredient in the extraction medium. For quantitative estimation, the supplied fungicide must be known. Comparison of inhibition zone size with a standard series allows determination of the approximate overall dosage of fungicide in the seed sample. The method is very useful for qualitative detection. By this method it is possible to measure the dosage on individual seeds by extracting chemical from each seed in the sample separately. However, not all fungicides are extracted equally well (Crosier *et. al.*, 1968).

The sensitivity of the method can be increased by using bi-directional diffusion of fungicide in agar bands inoculated with a test organism. By using approximately 1 mm thick and 10 mm broad agar bands, sensitivity of the test was increased by 8 fold compared to 1.6 mm thick agar plate (Edgington *et. al.*, 1973).

Chemical analysis

To carry through chemical analysis for qualitative or quantitative estimation of supplied fungicide, specially trained staff, special equipment and sophisticated methods are needed. This procedure is too costly to carry through as a routine test (Radtke, 1982), particularly in plant quarantine laboratories

Bio-assay

Amongst the methods proposed to detect presence and amount of chemical adhering to the seed surface, a bio-assay method has been claimed to be the most effective and efficient (Purdy, 1967). The most widely used bio-assay test for testing the quality of a seed treatment, is the test where inhibition of growth of a specific microorganism is noted using agar media.

The two fundamental principles of the bio-assay are diffusion of the chemical directly from the treated seed into an agar medium and growth inhibition of a sensitive microorganism in the medium due to the diffusing chemical (Ehle, 1973). Treated seeds are placed on solid or solidifying agar medium in containers, e.g. petri dishes, previously inoculated with a microorganism which is sensitive to the chemical on the seed surface. The compound starts diffusing from the seed surface in contact with the agar into the agar medium in all directions. The petri dishes are incubated in conditions which ensure optimum growth of the sensitive microorganism in the medium and optimum diffusion of the chemical. During the incubation period, the growth of the microorganism is restricted or impossible around the treated seed due to the action of the chemical. In this way a clear halo, the inhibition zone, is produced around each treated seed.

A suitable bio-assay test-organism should fulfill the following requirements:

- High sensitivity to the tested chemical or if the chemical is unknown, high sensitivity to a wide range of chemicals
- Rapid and uniform growth in a solid assay medium and
- Easy cultivation and abundant formation of viable propagules.

Viable propagules are, in the case of fungi, most often conidia. In addition to the above mentioned points, the test-organism should be insensitive to exudates from the untreated seed in the bio-assay. It is showed inhibition zones around untreated seeds of peanut, using *Curvularia* sp. as test-organism. Furthermore, a test-organism must be able to resist heat at about 35-40°C if it is mixed with the melted agar medium before pouring into petri dishes. Table 2 shows examples of different organisms used in the bio-assay for different fungicides.

If certain precautions are made, the size of the inhibition zone can be related to the quantity of chemical on the seed (Ehle, 1973) by measuring the diameter of the individual halos. The quantity of fungicide on seeds to be tested can be estimated by careful comparison of size of inhibition zones produced around the seed to the size of inhibition zones produced around seeds treated in the laboratory with known amount of chemical. When the bio-assay method is used qualitatively, reproducibility of the halo size is unimportant, but for a quantitative estimate of the applied fungicide, the reproducibility of the halo size is essential for the reliability of the method. The size of the inhibition zone is entirely dependent on the diffusion of the compound and the growth of the test-organism in the agar (Ehle, 1973). All factors affecting one or both, affect the size of the halos. The factors involved are complex and they are briefly presented below.

Agar concentration. Usually the assay medium contains agar in a concentration between 1 and 2 per cent (Ehle, 1973). By decreasing the agar concentration from 2 per cent to 1.25 per cent, the inhibition zone showed about 85 per cent increase in diameter (Jørgensen, 1966). The increase in inhibition zone size may be due to the more easy diffusibility of the chemical in a more liquid phase and to the fact that the plated seeds sink deeper into the more soft medium (Ehle, 1973). Less than 0.8 per - cent agar in the assay medium should be avoided because of incomplete solidification and likewise, more than 2 per cent agar should be avoided, because of the tendency to split when or if seeds are pressed into the agar.

Agar depth. The more shallow the medium is, the bigger the zone of inhibition (Ehle, 1973). This may be due to less loss of active ingredients towards the bottom in the shallow medium.

Crosier (1963) recommended to use an agar depth of 1.5 mm as a standard instead of depths of 2.5-4.5 mm. Halfon-Meiri and Dishon (1964) found that 1.5 mm agar depth was the minimum satisfactory depth for bio-assay of peanut seeds. However, using barley seeds and *Clavibacter rathay* as a test organism, Jørgensen (1966) observed biggest zones with 1 mm depth, with least variation.

Depth of sowing. Only the chemical in direct contact with the assay medium will diffuse into the medium. Crosier *et. al.* (1968) showed that the zone diameter increases with depth of sowing, but the relative zone diameter corresponding to the amount of fungicide on the seeds is unchanged. Therefore it is important that sowing of each seed in a test should be uniform to get comparable inhibition zones.

Age, distribution and amount of test-organism. Use of approximately the same number of viable propagules of the test-organism in different tests is essential for obtaining reliable and reproducible results. It is, therefore, important to transfer the test-organism to new media at regular intervals and to use only actively growing subcultures.

Even distribution of the test-organism in the medium is equally important to make sure that clear zones are due to inhibition of growth rather than absence of the test-organism. Even distribution is best obtained by mixing the propagules with melted, cooled agar before it is poured into containers. The amount of inoculum should be adjusted to produce the largest and most distinct zones of inhibition (Ehle, 1973) which is the basis for proper measuring of the diameter. Confluent zones must be avoided. In tests using spores of the fungus *Glomerella cingulata*, it was shown that 4000 spore/ml was most sensitive to mercurial dressings. Higher concentration of spores resulted in smaller inhibition zones (Ehle, 1973). Using a bacterium as test-organism, Jørgensen (1966) pointed out that the inhibition zones decrease in size with increase in bacterial concentration, especially when low concentrations of the chemical is assayed. Kovacs (1967) claim that the concentration of bacterial cells in the medium will influence the diameter of the inhibition zone only if the concentration is extremely low or extremely high.

Period of incubation. Isely (1965) recommended 24 hours or 48 hours of incubation before measuring the inhibition zones. However, Jørgensen (1966) observed a reduction in the inhibition zone by increasing the incubation period from 24 to 48 hours. Whether 24 or 48 hours is required, it depend entirely on the speed by which the test-organism multiply or germinate in the areas where it is able to grow and distinct inhibition zones are produced.

Seed size. Using wheat seeds of different sizes treated with "Ceresan L", all at equal rates, Crosier *et. al.* (1968) showed that the inhibition zones were approximately proportional to the surface area of the seeds in contact with the medium. However, this relationship was not found in barley.

Damaged seeds. Seeds with exposed endosperm at the time of treatment produce relatively larger inhibition zones than intact seeds. The reason is that more fungicide is absorbed in the exposed endosperm during the treatment.

Water content of seeds. Jørgensen (1966) showed that the inhibition zones produced by barley seeds, treated with organo mercury fungicides decreased in diameter with increasing water content of the seed.

Type of formulation of the chemical for seed treatment. Jørgensen (1966) showed a difference in inhibition zone diameter depending on the type of formulation such as powder-, oil- and liquid-formulated organo mercury fungicides. The largest zone diameter was obtained with powder formulations.

Because of the many factors which influence the inhibition zone diameter, it has not yet been possible to standardize the bio-assay to an extend that the use of controls in each test can be avoided. Seeds carefully and newly treated with the fungicide for which the tested seeds are analyzed, have to be extensively used as controls in each test. Control treatment should include laboratory treatment with ¼, ½, full and double of recommended dosage of the fungicide.

Ehle (1973) and Isely (1965) recommend to test at least 50 seeds of larger size and 100 seeds of smaller size taken at random in a representative seed sample. A similar number of seeds should be used as controls.

Sensitivity of the bio-assay. If the bio-assay is carefully carried out, seeds showing a mean inhibition zone diameter of about ± 20 per cent compared to the controls treated with full dosage, may be considered correctly treated. In case the mean zone diameter produced in the tested seed is close to the lower tolerance limit, a new bio-assay should be conducted before the final conclusion is drawn.

If the calculated standard deviation on the zone diameters exceeds the approximate value of 5, uneven distribution of the fungicide on the seeds is indicated.

Health precautions

All fungicides used for seed treatment are poisonous not only to the fungi against which they are applied, but also poisonous in various degree to man. The fungicides are usually classified according to their toxicity to man.

The most frequent way of treating seed is by application of fungicide which is deposited on the surface of the seed. The crust of fungicide is frequently easily released into the surroundings as dust. This phenomenon may best be observed when treated seed is stored and transported in bags. In general chemicals applied in a sticky dust formulation release more dust than chemicals 'glued' to the seed surface through a coating technique. Dust is more readily detached from seed treated with a larger dosage. Particularly plant quarantine personnel is exposed to seed with large doses of chemicals.

Poisoning with chemicals from treated seed may take place through direct skin contact which may cause skin disease or poisoning through absorption over the skin. Poisoning may also take place through inhalation of dust, or through eating items contaminated with chemical.

The release of chemical dust from the surface of treated seed is particularly high during sampling and dividing procedures. The impact from the surfaces of machinery on the seed passing through the machine usually cause release of high volumes of dust. Therefore, no machinery like the Boerner divider should be used for reducing samples of treated seed to working samples but manual sampling methods like the spoon method or repeated halving should be employed. The release of dust of chemicals is reduced in this way. However, the work must be carried out in a fume board as some release of dust will take place during handling of the treated seed. Since handling of treated seed is an important part of the plant quarantine work it is important that the risk of poisoning is reduced to a minimum through the following minimum precautions:

- Inform employees about the risk
- Train employees in handling dangerous substances
- Confine the area of operations with treated seed
- Put up warning signs for others
- Handle treated seed at a location with negative pressure device
- Room ventilation
- Wear mask, gloves, and laboratory coat
- Wash hands, arms, and face after handling treated seed
- Do not eat, store food, or smoke in the handling area
- Wash the floor. Do not sweep it
- Pregnant women should not handle treated seed.

References

Allam, A.I., Sinclair, J.B. and Schilling, P.E. 1969. Laboratory and greenhouse evaluations of four systemic fungicides. *Phytopathology* **59**: 1659-1662.

Arny, D.L. 1952. The bio-assay of Ceresan M on treated oat kernels. *Phytopathology* **42**: 222-223.

Blair, I.D. 1950. Wheat seed testing with reference to covered smut disease, pre-emergence loss, and fungicidal dust coverage. *New Zealand J. Sci. Technol.* **32A**(4):1-21.

Byford, W.J. 1975. Observations on the toxic effects of organomercury treatments on sugar beet seed. *Ann. App. Biol.* **79**: 221-230.

Crosier, W.F. 1963. Detection of seed treatment subcommittee. *Proc. Assoc. Off. Seed Analysts* **53**: 21-22.

Crosier, W.F., Glenn, B. and Walls, W.E. 1968. Biological determination of the amount of mercury on seeds. *Proc. Assoc. Off. Seed Analysts* **58**: 37-45.

Edgington, L., Buchenaver, H. and Grossmann, F. 1973. Bio-assay and transcuticular movement of systemic fungicides. *Pestic. Sci.* **4**: 747-752.

Ehle, H. 1971. Mikrobiologisches Nachweisverfahren zur Verteilung von Beizmitteln - in besondere Feuchtbeizmitteln - auf Getreidesaatgut. *Nachrbl. Deut. Pflanzenschutzdienst* **23**: 33-39.

Ehle, H. 1973. Microbiological methods for detecting dressings on treated seeds. *Residue Reviews* **45**: 125-143.

Ellis, M.A. and Sinclair, J.B. 1976. Effect of benomyl field sprays on internally borne fungi, germination and emergence of late harvested soybean seeds. *Phytopathology* **66**: 680-682.

Halfon-Meiri, A. and Dishon, I. 1964. A biological method for testing the efficiency of peanut seed treatment. *Plant Dis. Rep.* **48**(11): 853-854.

Hampton, P.D. and Rennie, W.J. 1978. Biological tests for seeds, pp. 24-43. In: *Seed Treatment.* (ed. K.A. Jeffs). DIPAC monograph 2.

Isely, D. (ed.) 1965. Rules for testing seeds. *Proc. Ass. Off. Seed Analysts* **54**: 1-112.

Jackson, D. and Ballard, N. 1973. The distribution and adhesion of mercury containing treatments on cereal seed. pp. 333-340. In: *Proceedings of the 7th British Insecticide and Fungicide Conference.*

Jeffs, K.A. and Tuppen, J. 1978. The application of pesticides to seeds, pp. 10-23. In: *Seed Treatment.* (ed. K.A. Jeffs). DIPAC monograph 2.

Jørgensen, J. 1966. Experiments with a biogical assay method for qualitative determination of fungicides on seeds of cereals. *Tidsskr. Planteavl* **70**(2): 244-251.

Kovacs, G. 1963. Biologiske metoders anvendelse til påvisning af fungiciders tilstedeværelse og bestemmelse af deres kvantitative forhold på sædefrø. *Ugeskrift for Landmænd* **108**: 598.

Kovacs, G. 1967. Den biologiske metodes anvendelse til påvisning af fungicider på sædekorn. Nyere erfaringer. *Tidskrift for Planteavl* **71**: 392-395.

Leben, C. and Keitt, G.W. 1950. A bio-assay for tetramethylthivramdisulfide. *Phytopathology* **40**: 950-954.

Machacek, J.E. 1950. An agar-sheet method of testing the efficiency of seed treating machines. *Can. J. Res. Sect. C.* **28**: 739-744.

Martin, J. 1975. Zu mikrobiologischen untersuchungen der verteilung quecksilberfreier beizmittel auf getreidesaatsgut. *Nachrichtenblatt des Deutschen Pflanzenschutzdienstes* **27**: 87-91.

Mead, H.W. 1945. A biological method of detecting the presence of fungicides on seeds. *Sci. Agr.* **25**: 458-460.

Molinas, S. 1961. Method for detecting fungicides on grain. *Cereal Sci. Today* **6**: 84-86.

Pepper, E.M. and Claussen, K.A. 1963. A rapid bio-assay for the detection of seed pesticides. *Plant Dis. Rep.* **47**: 374-377.

Purdy, L.H. 1967. Application and use of soil and seed treatment fungicide, I-II. Acad. Press.

Radtke, von W. 1982. Freivillige Beizstellenkontrolle unter Berücksichtigung der Umstellung auf quecksilberfreie Getreidebeizmittel. *Gesunde Pflanzen* **34**: 176-186.

Richardson, L.T. 1966. Reversal of fungitoxicity of thiram by seed and root exudates. *Can. J. Bot.* **44**: 111-112.

Samra, A.S. 1956. Relative value and mode of action of some fungicides used as seed disinfectants and protectants. *Mededelingen van de Landbouwhogeschool te Wageningen, Nederland* **56**: 1-51.

Sethofer, V. 1946-47. Studie o ucinrosti suchych moridel proti fusarioseplisni sneze. *Ochrana Rostlin* **19-20**: 56-75.

Smith, A.N. and Crosier, W.F. 1966. A comparison of methods and microorganisms for as saying treated seeds. *Proc. Assoc. Off. Seed Analysts* **55**: 104.

Thapliyal, P.N. and Sinclair, J.B. 1971. Translocation of benomyl, carboxin and chloroneb in soybean seedlings. *Phytopathology* **61**: 1301-1302.

Wallen, V.R. and Hoffmann, I. 1959. Fungistatic activity of captan in pea seedlings after treatment of the seeds or roots of seedlings. *Phytopathology* **49**: 680-683.

Seed Health Testing Laboratory: Facilities, Staffing and Training

S.B. MATHUR and HENRIK JØRSKOV HANSEN

Danish Government Institute of Seed Pathology for Developing Countries, Ryvangs Allé 78, DK-2900 Hellerup, Denmark

Introduction

Seed health refers primarily to the presence or absence of disease-causing organisms, such as fungi, bacteria, viruses as well as animal pests, such as eelworms and insects. Physiological conditions, such as trace element deficiency, may also be involved. Apart from the part on trace element deficiency, this definition, as formulated by the International Seed Testing Association (1985) describes very well the type of work which is conducted at plant quarantine stations.

The main objective of quarantine laboratories is to make sure that no living pest of quarantine importance enters the country of import and at the same time the exporting countries must ensure that consignments for export follow the requirements of the importing countries. This can only be achieved if the quarantine laboratories are adequately equipped with proper seed health testing equipment and staffed with well-trained seed health testing personnel. This, however, is generally not the case in quarantine stations, especially in the developing countries. Most stations are housed in unsuitable buildings with hardly any testing equipment and the personnel is not acquainted with proper seed health testing. In this article, an attempt has been made to provide basic information on how a seed health testing laboratory should be set up in a fairly well-established plant quarantine station including a rough idea on the cost of testing equipment. An account is also given on training possibilities of scientists and laboratory technicians.

A seed health testing laboratory

General plan

It is difficult or even impossible to suggest a blue print of a plant quarantine station which can be used as a standard all over the world. This is because of the fact that building requirements will vary from place to place depending on the types of quarantine organisms which are expected to be tested, as well as the overall workload. However, keeping in mind the common function of quarantine stations, it is possible to formulate a blue print where the main activities can be performed with special reference to seed. Such a plan is presented in Fig. 1. It is important that functions shown in the figure as well as the rooms for the staff are included in the quarantine building.

Since the purpose of a plant quarantine station is to exclude pests and pathogens and to prevent their escape to the outside environment, seed health testing must be done in a restricted area. Details of the restricted area are shown in Fig. 2. It houses handling of high-risk quarantine organisms where propagules of exotic organisms might be released during testing and diagnoses. The sequence of rooms within the restricted area should be followed as far as possible since the whole set-up is designed to allow a phytosanitory traffic pattern for keeping the risk of cross contamination to a minimum.

The interior decoration of rooms and laboratories is not presented in this paper. However, a few comments are made on a few general technicalities. All laboratory bench surfaces must be resistant to chemical decontamination and easily cleaned. The air conditioning system of the entire restricted area should be of the split-type by which exchange of air from inside to outside is avoided. If ordinary air conditioners are installed, it is necessary that the exchanged air is passed through a microbiological filter for trapping propagules of quarantine pests and diseases.

Testing equipment and cost

Testing equipment required to test seeds for infections of different organisms is largely the same as used in general plant pathological laboratories engaged in diagnostic work. Items considered most common are: compound and stereoscopic microscopes, centrifuges, balances, autoclave, basic glassware and chemicals. Generally, the plant quarantine laboratories are supposed to possess such equipment as they have to handle plant material of various types. When the testing material includes seeds, there is a need of additional equipment and facilities. There is a need of seed sampling devices for preparing representative working samples, extended incubation area fitted with light (near ultra violet light (NUV) or a combination of NUV and fluorescent day light) and an increased use of petri dishes for incubating seeds. The quarantine greenhouse facilities must also be extended to accommodate all work involving growing of seedlings and plants.

If basic plant pathological equipment is lacking in a quarantine station and seed health testing has to be started, the approximate cost of such equipment is summarized in Table 1 based on purchase in Denmark. Prices may vary from country to country.

With this equipment seeds can be tested by a range of methods commonly used for detecting fungi, bacteria, viruses and nematodes.

Fig.1 General Floor Plan of a Post-Entry Quarantine Building showing areas dedicated to special function.

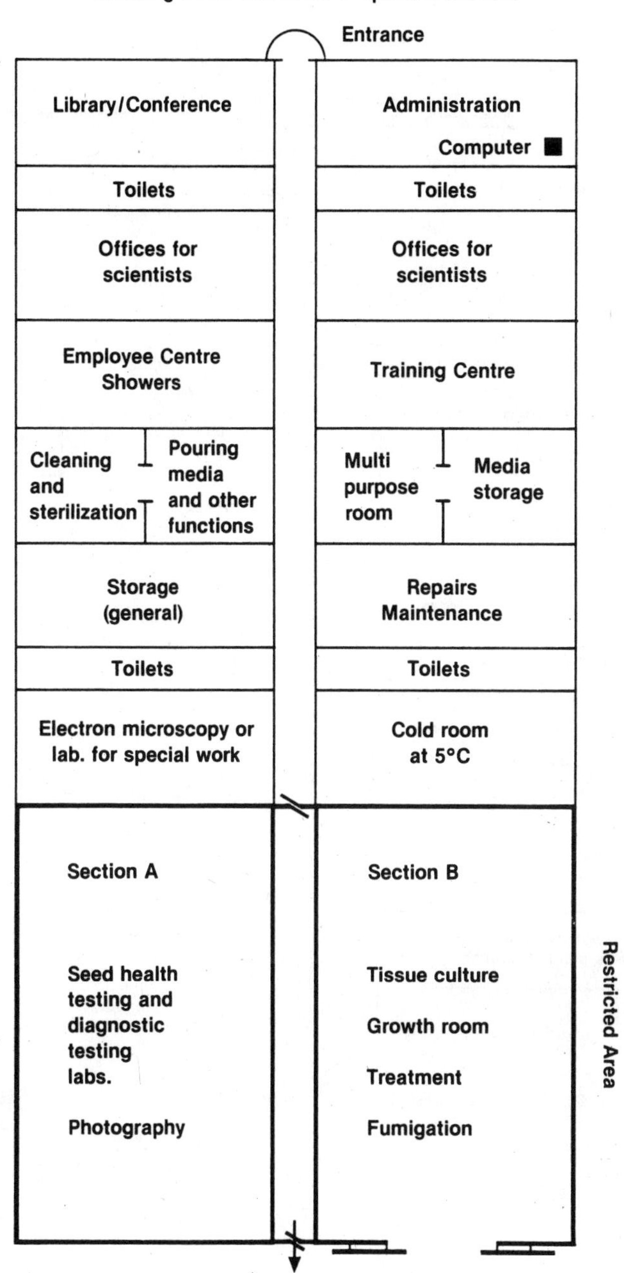

Fig.2 Restricted Area.

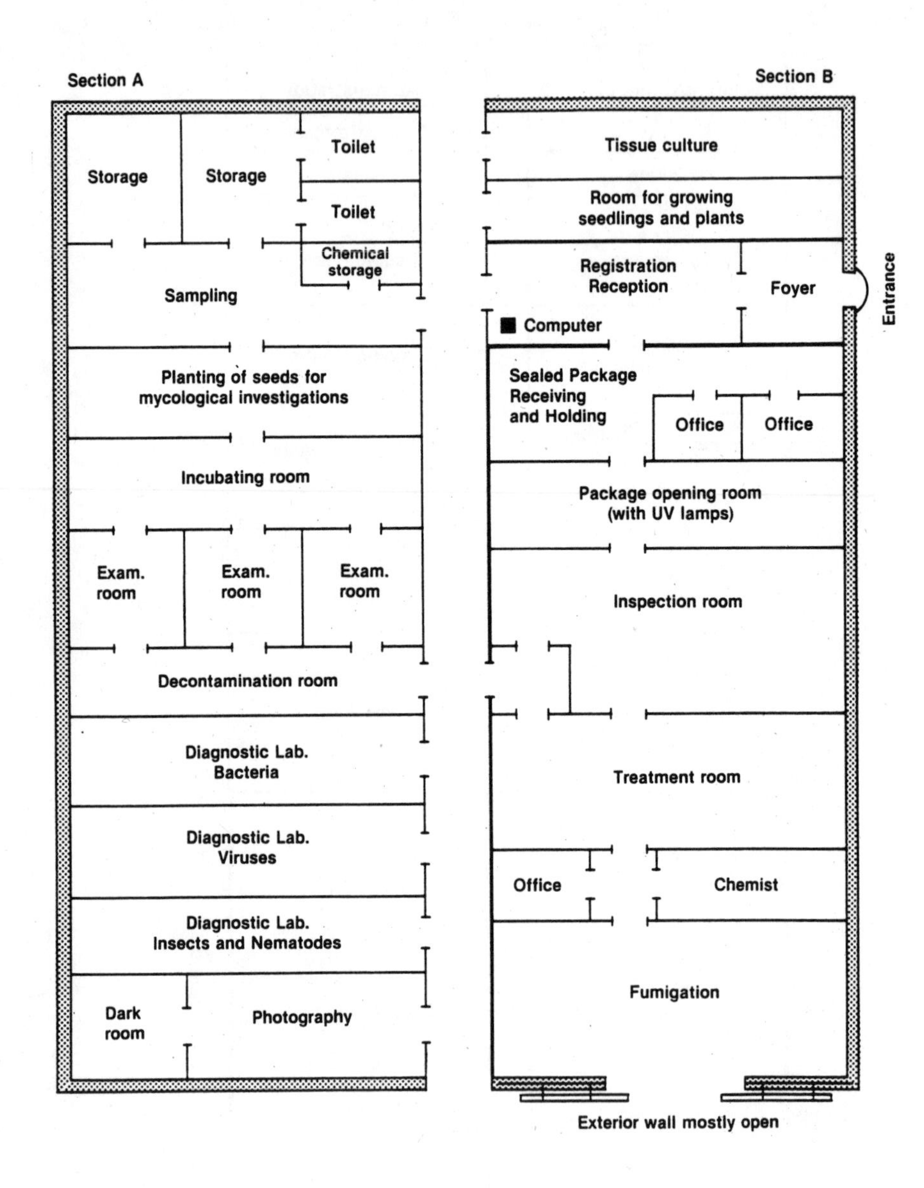

Table 1. Cost of laboratory equipment and consumables needed to establish seed health testing facilities in a quarantine station, based on purchase price in Denmark

	Danish Kroners (DKK)
Permanent items	
Sampling devices	25,000
Mycology	115,000
Bacteriology and virology	555,000
Nematology	10,000
	705,000
Consumables per year	
Mycology	5,000
Bacteriology and virology	50,000
Nematology	3,000
	58,000

Equipment lists are available at the Danish Government Institute of Seed Pathology for Developing Countries in Denmark on request.

Note: Cost of equipment required for bacteriology and virology is listed together as many items are shared.

Fungi

Inspection of dry seeds
Blotter method
Agar plate method
Embryo-count method
Seedling symptom test
Growing-on test

Viruses

Indicator plant test
Growing-on test
Microprecipitin test
Agargel diffusion test
Enzyme-linked immunosorbent assay (ELISA)
Dot immunobinding assay

Bacteria

Extraction and identification:
Direct plating of seeds on media
Liquid assay, seed washings
Liquid assay, ground-up seeds
Seedling symptom test
Biochemical test
Hypersensitivity test
Pathogenicity test
Serological test incl. IF microscopy

Nematodes

Inspection of dry seed
Baermann funnel method

Staff

The professional staff of a plant quarantine station should be able to process the amount of material which passes through the system and, at the same time, they must be competent enough to identify the organisms of quarantine significance. This can be achieved by appointing scientists specialized in mycology, bacteriology, virology, nematology, entomology, chemistry, horticulture and weed science. However, if funds or the quantity of material processed do not justify such an elaborate staffing appointment of a few professionals with a broad knowledge and experience would be advisable. The staff should have the possibility of inviting specialists located in other institutions of the country on consultation basis on short notice.

Training in seed health testing

Techniques used in testing seeds for the presence or absence of disease causing organisms are basically the same as those used in general plant pathological laboratories for isolating pathogens from diseased plant tissues. Scientists and technicians of quarantine can thus go to a variety of research laboratories where they can be introduced to a range of testing methods. There are, however, two institutions, one in Denmark and the other in Malaysia where specially tailored courses in seed health testing as well as in the principles of seed pathology are imparted each year. These courses are briefly described here.

Danish Government Institute of Seed Pathology for Developing Countries (DGISP)

Training in Denmark

It is important that scientists and laboratory technicians have a good understanding of the role seed-borne pathogens play in the field, their location in and on the seeds, in seed-transport containers, the different ways they attack seeds, seedlings and plants including losses caused, seed health testing procedures, and effective measures for their control at the seed and field levels. A thorough understanding of these aspects can be achieved through a systematic educational programme such as the one which was started in 1967 at the Danish Government Institute of Seed Pathology for Developing Countries (DGISP).

Scientists and technicians working at teaching and research institutions, in seed production and seed certification agencies, as well as in plant protection and plant quarantine, are admitted to the course organized in Denmark. So far, DGISP has trained 409 scientists and technologists from 67 developing countries in Denmark. Out of these, 54 have been trained from plant quarantine laboratories of 17 countries (Table 2). These numbers will increase to 79 and 22, respectively, if we consider those employees from other organizations like plant protection who are entrusted with the quarantine work. Such is the case of a number of trainees who came to our Institute from Egypt (11), Somalia (4), Sudan (3), Bangladesh (2), Nepal (5), Jamaica (2).

The course in Denmark is of nine months' duration divided into five major disciplines, general seed pathology, seed mycology, seed bacteriology, seed virology and seed testing. One of the important topics covered under general seed pathology is "Quarantine for Seed" where discussions are held on international spread of seed-borne diseases due to export and import of seed, interception of seed-borne diseases, principles of setting up quarantine regulations, principles of applying quarantine methods, pre- and post-entry control, and practice in adopting and applying quarantine provisions for seed. During practicals, the trainees test a variety of seeds for fungi, bacteria and viruses by appropriate testing methods. The trainees are trained in routine seed health testing to such an extent that they are able to conduct seed health testing in a rather efficient manner. The staff from quarantine organizations is especially in an advantageous position in our courses because they are able to test seed samples of different crops, originating from different countries of the world. Experience gained by examining seed samples with different infections is essential for officers and technicians working in quarantine laboratories.

Table 2. Number of plant quarantine scientists and technicians who have been trained in seed pathology and seed health testing, 1967 - 1991.

Country	Number	Year of training
Africa		
Ghana	1	1975
Kenya	5	1975, 76, 78, 82, 91
Madagascar	3	1980, 80, 88
Mauritius	5	1976, 82, 83, 86, 88
Mozambique	1	1985
Nigeria	6	1972, 73, 78, 85, 86, 86
Sudan	1	1990
Asia		
Burma	1	1985
People's Republic of China	5	1982, 83, 83, 83, 85
India	12	1967, 68, 70, 75, 76, 77, 78, 80, 80, 81, 81, 84
Indonesia	5	1975, 75, 84, 85, 85
Pakistan	2	1972, 72
Philippines	1	1986
South Korea	1	1979
Thailand	2	1968, 82
South America		
Cuba	2	1982, 82
Suriname	1	1976
17 countries	54	

The home institutions of our trainees are entitled to an equipment grant which amounts to 45,000 Danish Kroner (ca. 6,500 US$). The equipment is shipped to the recipient countries after the trainees have returned to work. In this way, seed health testing facilities have been created in a number of countries where seeds can be tested, especially for fungi and more recently also for bacteria and viruses.

Training in developing countries

The Institute of Seed Pathology has so far organized 15 short training courses, workshops and seminars. In almost all courses, participants from plant quarantine laboratories were present. Two of the courses were specially designed for personnel of plant quarantine, one held in India in 1980 for 26 participants from 24 countries of Africa and Asia and the second for 20 quarantine inspectors of Indonesia in 1988.

In 1989 the Institute organized a training course in seed health testing techniques for 20 technicians and scientists of Nepal in Kathmandu. Five of the participants were plant quarantine officers. The Nepal course was different from other courses as all the equipment which was taken from Denmark to run the course on seed-borne fungal, bacterial and viral diseases was given to the Central Division of Plant Pathology, National Agricultural Research and Services Centre as a gift. The Division is now fully equipped with a seed pathology laboratory where seed can be tested for health.

Another course which deserves special comments is the First International Course on Seed Health Testing of Rice organized by the Institute in 1988 at the International Rice Research Institute (IRRI) in the Philippines. Almost all the 19 participants were from quarantine and plant protection departments and they were trained for 6 weeks in testing techniques for detecting dangerous fungi and bacteria in rice seed. The main objective of the course was to make sure that the participants would be able to test rice germplasm for quarantine organisms, thereby making the movement of germplasm safer.

Response

Training in Denmark and abroad has resulted in introduction of seed health testing in plant quarantine laboratories of a number of developing countries. We will briefly describe here the work carried out and the achievements made in a few countries.

Seed health testing has become a routine work at the National Bureau of Plant Genetic Resources (NBPGR) in India, (Lambat *et al.*, 1983). Four staff members of the Bureau had obtained certificates in Seed Pathology at the DGISP in Denmark. Every year, the Bureau receives 60,000 to 100,000 samples of plant material of which 95-98% is in the form of true seed. The material under exchange is examined for dangerous organisms using internationally approved testing methods. Several important disease producing organisms, such as *Ustilago tritici, Urocystis tritici, Peronospora manshurica, Puccinia carthami, Fusarium nivale, Heterodera schachtii, H. goettingiana, Rhadinaphelenchus cocophilus, Oscinella frit, Acathoscelides obtectus, Bruchophagus glycyrrhizae, Ephistia elutella* and *Carpocapsa pomonella* have been intercepted. As far as possible, efforts are made to salvage the infected/infested material by using mechanical, chemical and physical methods. During the CTA Seminar on Seed Pathology held in Denmark in 1988 it was reported by Ram Nath of NBPGR that 98% of the material received was cleared and made available to scientists in pest and pathogen-free conditions.

Scientists trained in Denmark were instrumental in developing testing facilities at NBPGR and according to our information they have participated actively in the planning of the plant quarantine station at Hyderabad, which will process all germplasm imports of the International Crops Research Institute for the Semi-Arid Tropics (ICRISAT) and the All India Co-ordinated Rice Improvement Project.

Plant quarantine stations situated in Bombay, Calcutta and Madras have been assisted in conducting routine seed health testing work by Indian scientists trained in Denmark.

Seed health testing at the Plant Quarantine Service at Moor Plantation, Ibadan in Nigeria was started by scientists and technicians trained at DGISP. The initial lay-out of the laboratory was worked out by Dr M.O. Aluko, the Director of the Plant Quarantine Service. The blueprint of the Nigerian closed quarantine laboratories, including the seed health testing complex and details of post-entry glasshouses, have been presented by him in an article published in 1983. This comprehensive account with photographs and illustrations can be of immense use to those who wish to start similar work. The quarantine service is able to process over 70% of all plant importations by laboratory seed health testing, leaving only about 30% for growing-on tests in the closed quarantine glasshouses.

Implementation of routine seed health testing became inevitable as Nigeria witnessed the unfortunate introduction of *Xanthomonas campestris* pv. *manihotis*, the causal bacterium of bacterial blight of cassava, and two new pests, cassava green spider mite and mealy-bug. The cassava green spider mite (*Mononychellus tanajoa*) which was reported in Uganda in the early 1970s was found massively infesting cassava plots in Nigeria by 1979 while the mealy-bug (*Phenacoccus manihotis*), hitherto found only in Brazil, had by 1980 multiplied in some parts of Nigeria causing upto 70% reduction in Cassava production (Aluko, 1983). The losses suffered by the nation as a result of the disease have been recorded to be above two million Naira.

Olembo (1983) reported that due to the lack of trained personnel and facilities it became necessary in Kenya to cut down on processing seeds through quarantine and to increase those that come with permit through certified sources. In spite of this arrangement, pathogens such as *Xanthomonas campestris* pv. *vesicatoria, X. campestris* pv. *campestris, Phoma lingam, Sclerospora graminicola, Sclerophthora macrospora* on seeds of cereals were introduced into Kenya. The seed health testing work at the plant quarantine station of Muguga is being carried out by personnel trained in Denmark.

Indonesia and Thailand are two other countries worth mentioning. Seed health testing, particularly in Jakarta and Bangkok, has become a daily routine especially for detecting fungi in seeds. Some seed lots are now also tested for viruses.

Considerable difficulties have been encountered in the international transfer of plant genetic material due to quarantine regulations and practices. In June, 1980, late Dr Paul Neergaard and the senior author of this article, convened a meeting at the Institute of Seed Pathology on the dangers involved in indiscriminate movement of seed by the International Agricultural Research Centres (IARCs). Dr Joseph Karpati, the then FAO Plant Quarantine Officer in Rome at that time, and late Dr P.N. Nirula of ICRISAT were invited. Based on our recommendations Dr Robert Kahn of the USA was sent, as a FAO consultant, to most IARCs to assess the risks involved in exchange of germplasm. His report was presented in a meeting "International Consultation on a System for Safe and Efficient Movement of Germplasm" held at the Centro International de Agricultura Tropical (CIAT), Colombia in June 1982. As a result, routine seed health testing has been introduced at most of the IARCs in subsequent years. In addition, our Institute has taken part in training scientists at IRRI and ICRISAT. IRRI has one of the most well-equipped seed pathology laboratories of the world.

Asean Plant Quarantine Centre and Training Institute (PLANTI)

PLANTI was created in 1980 under the aegis of the ASEAN Committee on Food, Agriculture and Forestry. It has been funded by the United States Agency for International Development and the Government of Malaysia. Since its creation PLANTI has given ASEAN agriculture a new impetus by providing the region a frontline defence against invading plant pests. It has become the region's leading institution in plant quarantine with the objective to complement the PQ activities of the ASEAN member countries (Brunei Darussalam, Indonesia, Malaysia, the Philippines, Thailand and Singapore) and to provide a focal point and coordinating mechanism for improving these activities through training, research, information exchange and consultancy/advisory services.

Table 3. Training in seed-borne diseases and seed health testing at PLANTI (1981 - May 1991)

Training Course	No. of Courses	Duration of Course	No. of Participants
Short Course specializing in seed-borne diseases and seed health testing	3	3 weeks	53
Short courses incorporating seed-borne diseases and seed health testing as part of course since 1981	14	2-5 weeks	257
Certificate/Diploma since 1983	8	5 months	113

Training in seed health testing at PLANTI began in 1981 as part of a general plant quarantine course. Since then training of PQ officers in seed health testing has been carried out either as specialized courses with subject areas related totally to seed-borne diseases or general courses on disease/detection techniques.

The number of participants trained in seed-borne diseases and seed health testing is given in Table 3. Some of them had earlier received training in seed pathology in Denmark. These scientists are now manning PQ laboratories where they test incoming and outgoing seeds using the techniques acquired in Denmark and Malaysia.

References

Aluko, M.O. 1983. Seed health testing for quarantine in Nigeria. *Seed Science and Technology* **11**: 1239-1248.

International Seed Testing Association. 1985. International Rules for Seed Testing 1985. *Seed Science and Technology* **13**: No 2. 520 pp.

Lambat, A.K., Wadhi, S.R. and Sanwal, K.C. 1983. Seed health testing at the National Bureau of Plant Genetic Resources, India. *Seed Science and Technology* **11**: 1249-1257.

Olembo, S. 1983. Seed health testing at the plant quarantine station at Muguga, Kenya. *Seed Science and Technology* **11**: 1217-1223.

Control and Special topic

Physical and Chemical Seed Treatments

EDWIN FELIU

FAO, Via delle Terme di Caracalla, 00100 Rome, Italy

Introduction

Seeds and other propagative plant materials are known to be potential pathways for the introduction and spread of plant pests. Numerous insects, nematodes, fungi, bacteria and viruses are known to be seed-borne. In addition, mites and weeds are also associated with seeds.

Plant Quarantine treatments are aimed at either preventing or reducing the threat of introducing plant pests with consignments of seeds and other plant materials while causing minimal or negligible adverse effect on the host commodity.

This paper presents the most commonly used chemical and physical seed treatments for quarantine purpose. It must be emphasized that complete elimination of the concerned pests is not always possible.

Chemical seed treatments

Chemical seed treatments includes a variety of applications of toxic chemical compounds including: fumigation, sprays, dusts, dips and aerosols. Fumigation, dusts and dips are the most frequently used to treat seeds.

Fumigation

Fumigation is a chemical treatment that reaches the commodity primarily in the gaseous state. The most widely used fumigant in plant quarantine is methyl bromide. At present this fumigant has been employed in the treatment of a wide spectrum of plants and plant products, including seeds. The gas is applied to a commodity according to a prescribed schedule which is adjusted according to temperature and exposure period. It must be carried out in enclosures that will retain the gas with minimal loss during treatments. Stationary chambers are preferred, although tarpaulin tents made of plastic or canvas are widely used, particularly when fumigating large loads. Cargo containers, railroad cars, warehouses and ships holds have also been used, when appropriate. Methyl bromide is a preferred fumigant due to its excellent penetration, quick dispersal, plant tolerance in general and high insect toxicity. However, it is a highly toxic chemical which must be handled in a safe manner to prevent injury to users or damage to plants. Vacuum fumigation is conducted in gas-tight chambers capable of withstanding the reduced atmosphere. Most treatments require either a 660 mm or 380 mm vacuum.

Methyl bromide may be phytotoxic to certain plant species, and other products such as rubber and leather goods. With seeds there might be a reduction of the germination percentage.

Other fumigants used in plant quarantine include:

- Phosphine (Hydrogen phosphide) - It is used mainly to control insects infesting grains and stored products. Some insects such as *Tribolium castaneum* may develop resistance to this chemical.

- Hydrogen cyanide - It is highly toxic to humans and phytotoxic to plants, especially in the presence of moisture. For these reasons it has been replaced by other fumigants.

- Ethylene oxide - It is mainly used as a fumigant for insects in bulk grain and tobacco. Its use for seeds has been mainly as a surface sterilant and in seed devitalization.

Equipment generally used are:

a. Vacuum chambers: usually welded steel and reinforced with steel ribs; preferably rectangular, with doors at one end, hinged at sides; high-capacity vacuum pump; volatilizer, circulation system, fans, non-sparking explosion-proof type.

b. Tarpaulins: gas-proof, polyethylene, clear 6 mil thickness for fumigation under temporary structures.

c. Fumiscopes: portable with dry cells.
To measure concentration of fumigants. Air sample is pumped to a sensing element (in ounces/cu feet)

d. Polyethylene tubing - 7-8 mm internal diametre.

e. Fans: 30-40 cm diameter, 1500-2500 CFM non-sparking, induction type.

Dusts and dips

Dusting and dipping with fungicides and bactericides are the most widely used methods of seed treatment, especially against disease producing organisms. Many of them are only partially effective against specific organisms and may, thus, not be considered as full quarantine treatments. However, they may provide quarantine safeguards, particularly in eliminating or reducing surface infection of seeds. Some examples of specific chemicals used as dusts or dips include:

- Formaldehyde. It is used for dipping seeds and other plant materials. Skin contact should be avoided.

- Fungicides. Fungicides such as zineb, captan and ferbam are used in dips and dust forms as surface seed protectants.

- Sodium hypochlorite. It is used as a surface sterilant to eliminate nematodes, fungi and bacteria from seeds and other plant materials.

- 8 - Hydroxyquinoline sulphate. It is effective against surface-borne fungi and bacteria and causes no damage to seed viability. It is used for dipping seeds and other planting materials, especially citrus.

- Thiram. It is used as a seed protectant against a wide range of seed and soil-borne organisms.

Dipping tanks

Stainless steel tanks with individual compartments of different volumes for pesticide dipping and dusting are used.

Some examples of chemical treatments are:

Verticillium wilt of alfalfa. Dust with Thiram (tetramethylthiuram disulphide 80% a.i.) at 5 g per kg of seed. (Australia).

Citrus canker (*Xanthomonas campestris* pv. *citri*). Immerse in a solution of streptomycin (0.5 g streptomycin sulphate /L) for 30 minutes (Australia). Immerse in a freshly prepared solution of 8 - hydroxy quinoline sulphate (lg/O.lL) for 45 seconds (Australia). Immerse in a solution of 50% hydrogen peroxide for 10 minutes (S. Africa).

Marasmiellus cocophilus in partially husked coconut. Methyl bromide at normal atmospheric pressure (21 C + 32 g - 4 h), followed by dip in Captan or Thiram. Dip in 250 ppm PMA (phenyl mercury acetate) (Jackson, pers. comm).

Coffee berry (*Colletotrichum coffeanum*). Dust with Captan or Thiram then plant in post-entry quarantine.

***Dothistroma*, *Diplodia* and rusts of conifers.** MB at NAP 21 C + 32 g - 3 h plus immersion in 10 sodium hypochlorite solution containing 1 available chlorine for 10 min at 20 C. Auracaria seed may be frozen at -18oC for 7 days in lieu of fumigation (Australia).

Ascochyta blight & Fusarium wilt of chickpea. Dust with benlate.

Fungal and bacterial wilts of cotton. Treatment B.S followed by delinting in concentrated sulphuric acid, dry and dust with protective fungicides.

South American leaf blight (*Microcylus ulei*) of rubber. At origin: dip in formalin/water (1:100) for 15 min, dry and dust seed and packing material with 1% Daconil w/w (a.i. chlorothalonil).

At intermediate quarantine station: destroy containers and treat seed with Thiram or Agrocide (a.i. lindane) and repack using new packing material.

At destination post-entry quarantine: destroy container, dip seed in 0.1% mercuric chloride, dry, dust with Thiram or Agrocide (2% w/w) or Fernasan D (a.i. 25% Thiram and 20% BHC) and sow in post-entry quarantine.

Alternaria seedling blights of linseed. Dust with excess hydrated lime, place in 1 hydrated lime solution at 52°C for 10 min. Wash and dry. (Australia).

Downy mildew of Pearl millet. Dust with Metalaxyl.

Peanut rust (*Puccinia arachidis*) of peanuts. Dust with Thiram at 4 g per kg seed.

Fusarium wilt of safflower. Dust with Phygon (dichlone) or benlate.

Blister blight (*Exobasidium vexans*) of tea. Soak seed for 3 h in suspension of copper oxide or copper oxychloride in water @ 1 kg/150 litres (for surface contamination).

Treatment for bacterial diseases of tomato. Soak for 5 min in acidified mercuric chloride (l g mercuric chloride in 2.5 ml concentrated hydrochloric acid added to 2 litre of water) then dip in skimmed milk solution to neutralize.

Downy mildew and smuts of maize and sweet corn. Dry seed at 40°C for 24 h then either (a) soak seed in 3% sodium hypochloride solution for 15 minutes or (b) soak seed in 0.1% mercuric chloride for 10 min then wash in water and dry. Dust with Metalaxyl.

Controlled or modified atmospheres

This involves the manipulation of the levels of oxygen, carbon dioxide and nitrogen for the control of insects in stored grains and nut crops. At present, there are no quarantine treatments in use but the procedure is being investigated as a potential treatment.

Physical seed treatments

These include the use of non-chemical agents including thermotherapy (heat or cold) and irradiation. These treatments offer the following advantages over chemical treatments:

- These have wide application against a broad spectrum of plant pests, including insects, nematodes, mites, molluscs, fungi, bacteria and others.
- They leave no chemical residue on the treated commodities.
- They are safe to the user and to the environment.
- They are usually less phytotoxic.
- They are competitive in cost with chemical treatments.
- They are relatively easy to apply.

Some adverse effects, however, have been reported in treated plant commodities.

Thermotherapy

It involves the use of controlled cold or warm temperatures to eliminate plant pests by killing them.

a. Cold temperatures are used against insects, particularly fruit flies infesting fruit crops. Temperature range varies from 0°-2.2°C for periods of 10-22 days, depending on the type of host/pest combination involved. Treatments may be performed in refrigerated vans, warehouses or aboard vessels. Not used for seeds.

b. Use of sustained above ambient temperatures.

Dry heat. In quarantine this is used to treat soil and certain non-propagative plant materials. Recently, it has been used to treat fruits against fruit flies.

Dry heat ovens. Convection type with thermometer. Temperature Range 0 - 250°C.

Hot water. Hot water treatment has been used to treat seeds against fungi and bacteria. It has the following advantages:

- no toxic hazards or residues during use
- no pest resistance problems
- effective against a wide range of pests

Its use, however, is limited to relatively small consignments (up to 1 kg or 1,000 seeds).

The treatment is applied using a hot water bath (10-15 litre capacity) equipped with a stirring unit to ensure uniform heat distribution with a maximum variation of +0.5°C. Seeds are suspended in cheese cloth in the hot water for a specified time.

Hot water treatments are also applied to bulbs, rhizomes and other underground portions of plants against nematodes at temperatures ranging from 39° to 52°C for periods usually varying from 30 to 240 minutes.

Recently, hot water treatments have been used to treat fruits against fruit flies.

Use of stainless steel tanks, equipped with a thermostatically controlled heating element to ensure that the temperature of the water remains constant during the duration of the hot water treatment, is common. Temperature range: above ambient to 100°C, 5-15 litre capacity, with stirring unit.

Some examples of hot water treatment are presented in Table 1.

Vapour heat. It involves the use of heated air which is saturated with water vapour for raising the temperature to a required point and maintaining it during a specified time. The latent heat released by the condensation of the vapour on the commodity raises the pulp temperature evenly and quickly, thus avoiding damage to the fruit. This method is used to treat fruits against fruit flies but not for seeds.

Irradiation

The use of ionizing energy such as gamma rays, X-rays and electrons on host commodities provides quarantine security. It is used against insects and other pests, particularly fruit flies in fruits and vegetables. Low doses (up to 150 GY) are effective in preventing adult emergence in fruit flies, and a minimum of 300 GY will prevent other insects and mites from becoming established in non-infested areas. Irradiation is effective against a broad spectrum of plant pests whilst having no adverse effects on most host commodities at the required doses. It leaves no residue on the plant materials treated and is cost-competitive with other treatments. It has not been used to treat seeds against pests, but research on this application may demonstrate the feasibility of its application.

Combined treatments

Various chemical and physical treatments may be combined to achieve maximum efficacy against determined pests. With seeds, combination of treatments involve the use of dips and dusts, as well as fumigation and hot water treatment.

For example, sorghum seeds are subjected to hot water treatment at 52°C for 10 minutes followed by dry seed dressing with Metalaxyl to control the seed-borne infection of downy mildew pathogen (*Peronosclerospora sorghi*). Similarly, against *Uromyces fabae*, rust of sugar beet, the seeds are treated with sulphuric acid for 5 minutes followed by hot water treatment at 50°C for 30 minutes.

Table 1. Information on some common hot water treatments.

Seed	Pathogen	Pre-soak hours	Hot water dip		Remark
		Time Hours	Deg C	Min	
Pepper	*Xanthomonas vescicatoria*		52	30	
Pepper	*Pellicularia filamentosa*		52	30	
Potato (true seed)	*Phomopsis vexans*		50	30	South Africa
Rice unblemished	Miscellaneous seed-borne diseases		57	15	Australia
discoloured			60	15	
Rye	*Ustilago* spp.	5	54	10	
Sesame			52	10	
Sunflower	*Plasmopora halstedii*	4	52	20	
Tobacco			55	30	
Tomato	*Rhizoctonia* spp.		52	30	
Triticale		4-5	52	10	Australia
Wheat	*Septoria nodorum* *Ustilago nuda*	4-5	54	10	Australia

Conclusions

Plant quarantine treatments for seed aim at eliminating plant pests with minimal adverse effect on the material treated. There are chemical and physical forms of treatments. The first group utilizes a toxic chemical while the second uses temperature or irradiation. Selection of treatments depends on the host/pest relationships involved. Chemical treatments are very effective against specific pests when used alone or in combination. However, they may cause phytotoxicity to certain seeds and plant materials, and may leave objectionable residues. Moreover, they may be hazardous to the user or the environment. Physical treatments are, in general, applicable to a wider spectrum of plant pests, leave no residues and are safe to apply. Some of them, however, may adversely affect germination.

Although a substantial number of seed treatments may provide quarantine security against various plant pests, many are only partially effective or ineffective in doing so. The ultimate choice in selecting a quarantine treatment for seed must take this fact into consideration.

References consulted

Anonumous. Irradiation as a Quarantine Treatment: A Report of the Task Force Convened by thr International Consultative Group on Food Irradiation (ICGFI). Bethesda, MD.

Anonymous. 1982. Report of the FAO/DANIDA Seminar on Quarantine for Seed for Developing Countries of Africa and Asia. GCP/INT/322/DEN. New Delhi, India.

Anonymous. 1991. *Plant Quarantine Treatment Manual.* USDA-PPQ-APHIS.

FAO. 1983. *International Plant Quarantine Treatment Manual.* FAO Plant Production and Protection Paper No. 50. Rome.

FAO. 1989. *Manual of Fumigation for Insect Control.* FAO Production and Protection Paper No 54. Rome.

FAO. 1991. *Glossary of Phytosanitary Terms.* Rome.

Hewitt, W.B. and Chiarappa, L. 1977. *Plant Health and Quarantine in International Transfer of Genetic Resources.* CRC Press Inc.

Kahn, R.P. 1989. *Plant Protection and Quarantine.* Vol. I. CRC Press. Boca Raton, Florida. 226 pp.

Kahn, R.P. (ed.). 1989. *Plant Protection and Quarantine.* Vol. II. CRC Press. Boca Raton, Florida. 265 pp.

Kahn, R.P. (ed.). 1989. *Plant protection and Quarantine.* Vol. III. CRC Press. Boca Raton, Florida. 215 pp.

Neergaard, P. 1980. A review on quarantine for seed. National Academy of Sciences, India. Golden Jubilee Commemoration Volume: 495-530.

Elimination of Infection from Valuable Seed Germplasm

P.E. KYRIAKOPOULOU

Benaki Phytopathological Institute, GR 145 61 Kiphissia-Athens, Greece

Introduction

Seed pathology is important (Chiarappa and Gambogi, 1986), because it is related to food production in various ways. Seed-borne diseases affect the production of food and feed, both in terms of quantity and quality. The pathogens can negatively affect seed viability and seedling vigour. Infected seeds act as primary infection *foci* randomly scattered within the field, right from the emergence of the seedlings and from whose *foci* the pathogen spreads, especially under favourable epidemiological conditions, and, through transportation, they contaminate previously disease-free areas, collections, breeding lines etc.

At least 1300 seed transmitted plant pathogens are known to infect a variety of plant species (Diekmann, 1990). Some examples are *Ascochyta* spp. in legumes, *Drechslera* (*Helminthosporium*) spp. in cereals, *Pyricularia oryzae* in rice, *Septoria* spp. in cereals and other crops, bunts (*Tilletia* spp.) and smuts (*Ustilago* spp.) in cereals, pathovars of *Pseudomonas syringae* and of *Xanthomonas campestris* in legumes and other crops, barley stripe mosaic virus in barley and other cereals and grasses, bean common mosaic virus in bean and cowpea, a large number of other legume viruses, prunus necrotic ringspot virus in *Prunus* spp., the nepo viruses in general, *Ditylenchus dipsaci* in faba bean and other hosts, etc. In Table 1 a list of fungal, bacterial, viral and nematode seed-borne pathogens of crops important for dry areas is presented.

Table 1. Seed-borne pathogens of cereals, legumes, sugarbeets, forage, oil seed and fiber crops

A. FUNGI (Compiled mainly from CMI Descriptions of Pathogenic Fungi and Bacteria, and Neergaard 1979. More examples in Agarwal and Sinclair, 1987).

Pathogen	Host
Alternaria brassicicola	rape
Alternaria sesami	sesame
Alternaria triticina	wheat

.....cont.

Pathogen	**Host**
Ascochyta fabae	faba bean
Ascochyta gossypii	cotton
Ascochyta lentis	lentil
Ascochyta pinodes	pea
Ascochyta pisi	pea
Ascochyta rabiei	chick pea
Ascochyta sorghi	sorghum
Aureobasidium lini	flax
Cercospora arachidicola	peanut
Cercospora beticola	beet
Cercospora personata	peanut
Cercospora sesami	sesame
Claviceps microcephala	sorghum millets(*Paspalum scrobiculatuum* and *Pennisetuum typoides*)
Claviceps paspali	millets (*Paspalum scrobiculatuun* and *Pennisetuum typhoides*)
Claviceps purpurea	rye
Cochliobolus sativus	wheat and barley
Colletotrichum gossypii	cotton
Colletotrichum lindemuthianum	bean
Colletotrichum lini	flax
Corynespora cassiicola	sesame, soybean
Diaporthe phaseolorum var. *caulivora*	soybean
Diaporthe phaseolorum var. *sojae*	soybean
Diplodia spp.	maize
Drechslera avenae	oat
Drechslera graminea	barley
Drechslera maydis	maize
Drechslera oryzae	rice
Drechslera sesami	sesame
Drechslera teres	barley
Drechslera turcica	maize, sorghum
Fusarium graminearum	maize, oat
Fusarium moniliforme	maize, sorghum
Fusarium nivale	oat, rye
Fusarium oxysporum f.sp. *ciceri*	chick pea
Fusarium oxysporum f. sp. *lentis*	lentil
Fusarium oxysporum f.sp. *lini*	flax
Fusarium oxysporum f.sp. *vasinfectum*	cotton
Fusarium oxysporum f.sp. *pisi*	pea
Fusarium oxysporum f.sp. *phaseoli*	bean
Fusarium spp.	barley, wheat
Gloeocercospora sorghi	sorghum
Glomerella gossypii	cotton
Melampsora lini	flax
Myrothecium roridum	soybean
Periconia circinata	sorghum
Peronospora manshurica	soybean
Phoma exigua var. *linicola*	flax
Phoma medicaginis var. *pinodella*	pea

.....cont.

Pathogen	Host
Physalospora rhodina	cotton, peanut
Plasmopara halstedii	sunflower
Pyricularia grisea	millet (*Panicum miliaceum*)
Pyricularia oryzae	rice
Pyricularia satariae	millet (*Eleusine coracana*)
Rhizoctonia soalni	cotton
Rhynchosporium secalis	barley
Sclerospora graminicola	millets (*Panicum miliaceum*, *Pennisetum typhoides*, *Setaria italica*)
Sclerotinia sorghi	sorghum
Sclerotinia sclerotiorum	bean, chick pea, sunflower
Septoria avenae	oat
Septoria avenae f.sp. *triticea*	wheat
Septoria linicola	flax
Septoria nodorum	wheat
Septoria tritici	wheat
Sphacelotheca cruenta	sorghum
Sphacelotheca destruens	millet (*Panicum miliaceum*)
Sphacelotheca sorghi	sorghum
Tilletia caries	wheat
Tilletia contraversa	wheat
Tilletia foetida	wheat
Tilletia indica	wheat
Tolyposporium ehrenbergii	sorghum
Tolyposporium indica	wheat
Ustilago avenae	oat
Ustilago crameri	millet (*Setaria italica*)
Ustilago hordei	barley, oat
Ustilago maydis	corn
Ustilago nigra	barley
Ustilago segetum f. sp. *hordei*	barley
Ustilago segetum f.sp. *tritici*	wheat
Urocystis occulta	rye
Urocystis agropyri	wheat
Uromyces betae	beet
Verticillium albo-atrum	cotton, sunflower

BACTERIA (Compiled mainly from CMI Descriptions of Pathogenic Fungi and Bacteria, and Neergard, 1979. More examples in Agarwal and Sinclair, 1987.)

Pathogen	Host
Curtobacterium flaccumfaciens pv. *flaccumfaciens*	bean
Clavibacter tritici	wheat
Erwinia stewartii	corn
Pseudomonas syringae pv. *sesami*	sesame
Pseudomonas syringae pv. *glycinea*	soybean
Pseudomonas syringae pv. *phaseolicola*	bean
Pseudomonas syringae pv. *pisi*	pea
Xanthomonas campestris pv. *glycines*	soybean
Xanthomonas campestris pv. *phaseoli*	bean

.....cont.

Pathogen	Host
Xanthomonas campestris pv. *sesami*	sesame
Xanthomonas campestris pv. *transluscens*	barley, wheat
Xanthomonas campestris pv. *holcicola*	sorghum
Xanthomonas campestris pv. *malvacearum*	cotton
Xanthomonas campestris pv. *oryzae*	rice

VIRUSES* (Compiled mainly from Bos, Hampton and Makkouk 1988; CMI/AAB Descriptions of Plant Viruses; Frison *et al.*, 1990; Hamilton 1983; Hampton 1983. More examples in Agarwal and Sinclair, 1987.)

* Some cases refer to experimental infections

Virus	Host and percentage of transmission
Alfalfa mosaic virus	alfalfa (*Medicago sativa*, up to 6%), chick pea, faba bean, lentil, *Medicago polymorpha* (0.2-49%), *M. trunctula* (2%), pea, other legumes
Apple mosaic virus	*Vigna* sp. (2%)
Arabis mosaic virus	beet (13%), soybean (6.3%)
Asparagus bean mosaic virus	*Vigna* sp. (3.5%)
Barley mosaic virus	barley (2-45%)
Barley stripe mosaic virus	barley (up to 100%), grass species (2-8%), oat (0-9.5%), wheat (6.7-81%)
Bean common mosaic virus	bean (*Phaseolus vulgaris* 7-83%), cowpea (*Vigna sequipedalis* 37%, *V. sinensis*, *V. unguiculata*, *Vigna* sp. 25-40%), *Phaseolus* spp.
Bean pod mottle virus	soybean (0.08%)
Southern bean mosaic virus	bean (*Phaseolus vulgaris*), cowpea (*Vigna sinensis* 3-4%, *V. unguiculata*)
Bean western mosaic virus (now considered strain of BCMV)	bean (*Phaseolus vulgaris*, 2-3%; *Phaseolus acutifolius* var. *latifolius* 7-22%)
Bean yellow mosaic virus	bean (*Phaseolus vulgaris*), cowpea (*Vigna sesquipedalis*), faba bean (0.1-24%), lupin (*Lupinus albus*, *L. luteus* 3-6.2%), *Melilotus alba* (3-5%), white sweet clover pea (5-30%)
Bean yellow mosaic virus, pea mosaic strain	pea (0.5%)
Beet 41 yellows virus	beet (47%)
Blackeye cowpea mosaic virus	cowpea (*Vigna unguiculata*, *Vigna* sp., up to 40%)
Blackgram leaf crinkle virus	cowpea (*Vigna mungo* 8%, *Vigna* spp. 20-42%)
Broad bean mosaic virus	broad bean
Broad bean mild mosaic virus	broad bean
Broad bean mottle virus	broad bean (1-2%), *Phaseolus mungo* (0-6.7%), *Vigna* spp. (6-7%)
Broad bean stain virus	bean (1-16%), broad bean (1-10%), lentil (14%), pea
Broad bean true mosaic virus	broad bean (1-15%)
Broad bean wilt virus	broad bean, lentil, pea, other legumes
Cherry leafroll virus	bean (12-40%), soybean (up to 100%)
Clover red vein mosaic virus	broad bean (up to 100%) other legumes
Clover (red) virus	*Trifolium pratense* (12-18%)
Clover (white) mosaic virus	*Cassia occidentalis*, *Trifolium pratense* (6%)
Clover (white) yellow mosaic virus	*Trifolium pratense* (7.6%)
Cocoa necrosis virus	*Phaseolus* sp. (1-24%)

.....cont.

Virus	Host and percentage of transmission
Cowpea aphid-borne mosaic virus	cowpea (*Vigna sinensis* 0.3-1.6%, *V. unguiculata* 3-19%, *Vigna* sp. up to 40%)
Cowpea banding mosaic virus	cowpea (*Vigna* sp. 15-31%)
Cowpea chlorotic spot virus	cowpea (*Vigna* sp. 3-16%)
Cowpea isometric virus	cowpea (*Vigna* sp. 5-16%)
Cowpea mild mosaic virus	soybean
Cowpea mild mottle virus	cowpea (*Vigna* sp. 5-90%), *Phaseolus* sp.(6%)
Cowpea mosaic virus	cowpea (*Vigna catjang* 17%, *V. sesquipedalis* 8%, *V. sinensis* 23%, *V. unguiculata* 1-5%)
Cowpea mosaic virus (aphid transmitted, isometric)	cowpea (*Vigna sinensis* 5-16%, *V. unguiculata*)
Cowpea mosaic virus (beetle transmitted)	cowpea (*Vigna sinensis* 10%, *V. unguiculata*)
Cowpea mottle virus	cowpea (*Vigna unguiculata* 0.2-10%, *Vigna* sp.
Cowpea ringspot virus	cowpea (*Vigna unguiculata* 10-30%)
Cowpea severe mosaic virus	cowpea (*Vigna sesquipedalis* 8%, *V. unguiculata* 0.2-10%)
Cowpea stunt	cowpea (*Vigna unguiculata*)
Crimson clover latent virus	crimson clover
Cucumber mosaic virus	bean (1-30%), *Phaseolus aureus* (0.3%), cowpea (*Vigna cylindrica, V. sesquipedalis, V. sinensis* 4-30%, *V. unguiculata* 15-20%), *Lupinus albus, L. angustifolius* (3-34%), *L. luteus* (21%), safflower
Echtes Ackerbohnenmosaik virus	broad bean (1-15%), cowpea (*Vigna* sp.1-15%)
Grapevine fanleaf virus	soybean (0-59%)
Guar symptomless virus	*Cyamopsis tetragonoloba* (12-70%)
Lima bean mosaic virus	*Phaseolus* sp. (25%)
Lima bean virus with mild mottle	*Phaseolus lunatus* (0.3%)
Lucerne Australian latent virus	alfalfa (0-8%)
Lucerne latent virus	alfalfa
Lucerne transient streak virus	*Melilotus albus* (2.5%)
Maize dwarf mosaic virus	maize
Maize leaf spot virus	maize
Maize mosaic virus	maize
Mungbean mosaic virus	*Phaseolus aureus* (0-30%), *Phaseolus* sp.(8-32%)
Oat mosaic virus	oat
Pea early browning virus	broad bean (1-10%), pea (1-37%)
Pea enation mosaic virus	pea
Pea false leafroll virus	pea (40%)
Pea leaf rolling mosaic virus	pea (2-55%), *Vicia* spp.
Pea mild mosaic virus	pea (15%)
Pea mosaic virus	lupin, pea, sweet pea, *Trifolium* spp., yellow lupin
Pea seed-borne mosaic virus (syn. pea fizzletop virus)	broad bean, lentil (32-44%), pea (0-90%), *Vicia* spp.
Peanut bunchy top virus	peanut
Peanut chlorosis	peanut
Peanut clump virus	peanut
Peanut marginal chlorosis virus	peanut (30-100%), peanut (6-14%)
Peanut mosaic virus	peanut

.....cont.

Virus	**Host and percentage of transmission**
Peanut mottle virus	peanut (0-20%), *Lupinus albus* (0.4%), bean (0-1%), cowpea (*Vigna uncuiculata* <1%)
Peanut ringspot virus	peanut
Peanut stripe virus	peanut (0.1-10%)
Peanut stunt virus	peanut (0.2%)
Raspberry ringspot virus	beet (50-55%), soybean (7.2%)
Runnerbean mosaic virus	*Phaseolus coccineus* (syn. *P. multiflorus* 42%)
Southern bean mosaic virus	*Phaseolus* sp. (1-30%), cowpea (*Vigna unguiculata* 1-40%, Vigna sp. 3-4%)
Soybean bud blight	soybean
Soybean mild mosaic virus	soybean
Soybean mosaic virus	soybean (0-68%)
Soybean stunt virus	soybean (50-95%), *Vigna* sp. (5%)
Subterranean clover mottle virus	*Trifolium subterraneum* (3%)
Sugarcane mosaic virus	maize (0.4%)
Sunflower mosaic virus	sunflower
Sunn-hemp mosaic virus	cowpea (*Vigna catjang* 17%, *V. unguiculata* 4-20%)
Tobacco mosaic virus	*Vigna* sp. (14%)
Tobacco ringspot virus	*Vigna* sp. (82%), soybean (1-100%)
Tobacco streak virus	soybean (0-90%), *Phaseolus* sp. (1-27%)
Tomato aspermy virus	bean (19%)
Tomato black ring virus	beet (56%), cowpea (*Vigna sinensis*) (23%), soybean (83%)
Tomato ringspot virus	soybean (76%)
Urdbean leaf crinkle virus	cowpea (*Vigna unguiculata* 6-15%, *Vigna* sp. 18-30%), *Phaseolus aureus, P. mungo*
Vicia cryptic virus	broad bean (88%)
Wheat streak mosaic virus	maize
Wheat striate mosaic virus	maize

VIROIDS

Several viroids are known to be true-seed borne, as potato spindle tuber viroid. But so far to my knowledge, viroids in the seed plants under consideration have not been reported.

NEMATODES (Compiled mainly from CIH Descriptions of Plant-Parasitic Nematodes, and Neergaard, 1979)

Anguina tritici	wheat
Aphelenchoides oryzae (*A. besseyi*)	rice
Ditylenchus angustus	rice
Ditylenchus dipsaci	broadbean

Recorded cases of valuable true seed germplasm infection

It is widely accepted internationally that use of pathogen-free germplasm is of vital importance in agricultural production. However, there are cases where important seed-borne pathogens have been encountered in germplasm collection around the world.

In 1966 the barley (*Hordeum vulgare*) breeding lines developed at Bozeman, Montana, USA, became infected by barley stripe mosaic virus (BSMV), presumably through pollen (Hockett *et al.*, 1988; Carrol *et al.*, 1990). BSMV is a virus which can be transmitted through seed, pollen, ovules and foliage contact. Infected seed when planted in the field nurseries is known to cause yield reduction of 25-46% (Carrol *et al.*, 1990).

The invaluable USA Plant Introduction Collection of pea (*Pisum sativum*) germplasm was contaminated with pea seed-borne mosaic virus, from infected accessions introduced mainly from India, which served as a major inoculum reservoir during the 1970's (Hampton and Braveman, 1979; Hampton, 1983). At least six North American institutional breeding programmes were involved in PSbMV outbreaks. Subsequent examination of the USA pea germplasm accessions disclosed that at least 420 out of 1835 accessions contained PSbMV.

A strain of PSbMV not infectious to peas was found to be seed transmitted in USA germplasm accessions of lentil (Hampton, 1982).

Bean common mosaic virus (BCMV), a potyvirus with highly virulent strains, ecologically similar to PSbMV, has been known for decades to occur as a seed-borne virus in *Phaseolus* germplasm collections. Current knowledge on BCMV strains (Drijfhout, 1978) and their incidence in BCMV epidemics suggest that special precautions should be taken to preclude dissemination of highly virulent BCMV strains in strategic *Phaseolus* seed stocks (Hampton, 1983).

In India a strain of BCMV was found to be seed-borne in urdbean (*Vigna mungo*) germplasm (Agarwal *et al.*, 1979).

Soybean mosaic virus found in soybean *(Glycine max)* germplasm was shown to significantly reduce the yield of selected germplasm accessions (Goodman and Oard, 1980).

A unique strain of cucumber mosaic virus readily transmitted to bean (*Phaseolus vulgaris*) was discovered in germplasm breeding lines in Idaho, Washington and Oregon, USA (Davis *et al.*, 1981).

Urdbean leaf crinkle virus was detected in Indian accessions of mungbean (*Vigna radiata*) germplasm (Beniwal *et al.*, 1980).

The risk of spreading of viruses with germplasm collections has been discussed by Bos in 1977.

Blight of chickpea (*Cicer arientinum*) caused by *Ascochyta rabiei* was found widespread and severe in research plots at the Regional Plant Introduction Station at Pullman, Washington, USA, in 1984. Eleven out of 17 chickpea accessions were susceptible to blight and the fungus was transmitted in 1-59% of the seeds (Kaiser and Hannan, 1988).

Kaiser and Hannan (1982, 1986) reported the incidence of *Ascochyta lentis* in lentil seeds imported to the USA.

Hampton (1983) suggests that maize dwarf mosaic virus and sugarcane mosaic virus in corn, soybean mosaic virus, tobacco ringspot virus and tobacco streak virus in soybean, and barley stripe mosaic virus in wheat are of major economical importance and their elimination should be given priority. Each disease outbreak involving either BCMV or PSbMV has been traced to infected seed. Both viruses in their respective hosts, therefore, offer opportunity for effective disease control through virus elimination from crop germplasm and nuclear seed sources (Hampton, 1983).

Countries unaffected by certain seed-borne pathogens or virulent strains of pathogens need to be protected from the introductions of such pathogens. For instance Syria should be protected from *Tilletia indica* (Karnal bunt), *Tilletia controversa* (dwarf bunt), *Septoria nodorum* (glume blotch) and other pathogens non endemic in the country (Diekman, 1990).

Elimination of infection

Considering the fundamental importance of germplasm it is obvious that it should be free from pathogens of economic significance.

In the commercial seed germplasm, due to its massive quantities, tolerance limits for seed infection are necessarily established, so that disease outbreaks are unlikely to occur. These tolerance limits, of course, are gradually diminishing down to zero, or close to zero, across the multiplication chain from certified to prebasic material. Lettuce seed in California should contain less than 1 infected seed in 30,000 (Grogan, 1983). Field and garden bean seed plantations in USA and Canada are allowed 1% mosaic-diseased plants if grown for certified, and 0.5% if for registered seed (Bos, 1977). Usual standards for loose smut (*Ustilago nuda*) in certification programmes are 0.1-0.2% infected embryos for the earliest three multiplication categories (Smith *et al.*, 1988).

In the case of valuable seed germplasm, tolerance for seed-borne pathogens needs to be zero. No seed-borne pathogens in any individual seed of this germplasm can be tolerated, because of its essential value, and because of its limited quantity making feasible the pathogen eradication task. So, valuable seed germplasm has to be or become of the elite (nuclear) status.

Pathogen elimination from seeds can be obtained in various ways. Methods or procedures used, often in combinations, are mechanical, physical, chemical, cultural or selective. Their use depends on the nature of the pathogen and the seed concerned, and on the available facilities. The efficiency also depends on whether the pathogen is present as a contamination in the seed lot, on the surface of the seed, and on its location within the seed as well as the volume of the seed to be treated. In Table 2 some examples are given on the location of pathogens in seeds.

Physical treatments

Removal of infected seeds. This is the most rudimentary method which is applicable in those cases where the seed lot consists of few seeds, and where the seed infection is externally obvious.

Wheat seeds affected by the nematode, *Anguina tririci*, called nematode galls or ear cockles, containing thousands of closely packed dehydrated bodies of nematode larvae, are dark and smaller than normal seeds. These galls are easily located and removed by hand. If the seed ot is large, affected seeds can be separated by sieving or by floating in 20% NaCl. Wheat seeds infected by *Tilletia* spp. are easily breakable, releasing black dust of the fungus teliospores. The infection of bean seeds by *Colletotrichum lindemuthianum* is usually expressed by red-brown lesions on the seed surface. Broad bean seeds infected by *Ditylenchus dipsaci* or broad bean stain virus show brown patches on their surface. Bean or soybean seeds with dull brown discoloration may be infected by *Pseudomonas syringae* pv. *phaseolicola* and infected seeds usually fluoresce under UV light. Partial elimination of barley stripe mosaic can be obtained by sieving out the smaller seeds; of bean yellow mosaic virus by discarding larger yellow lupine infected seeds, and of pea seed-borne mosaic virus by removing pea seeds with growth cracks (Bos, 1977).

Table 2. Location of pathogen in seeds (Data mainly from the book of Neergaard, 1979)

Embryo infection	
Viruses	Barley stripe mosaic virus (barley, wheat)
	Bean common mosaic virus (bean)
	Bean yellow mosaic virus (*Lupinus albus*)
	Cowpea aphid-borne mosaic virus (cowpea)
	Cowpea beetle-transmitted mosaic virus (cowpea)
	Lettuce mosaic virus (lettuce)
	Pea seed-borne mosaic virus (pea)
	Tobacco ringspot virus (soybean)
	Tomato black ring virus (soybean)
Fungi	*Ascochyta pisi* (pea)
	Fusarium culmorum (wheat)
	Fusarium moniliforme (rice)
	Ustilago tritici (barley, wheat)
Endosperm infection	
Viruses	Cucumber green mottle mosaic virus (cucumber)
	Tobacco mosaic virus (tomato mosaic virus) (tomato)
Fungi	*Sclerospora philippinensis* (maize)
Bacteria	*Erwinia stewartii* (maize)
	Clavibacter michiganensis subsp. *insidiosus* (alfalfa)

.....cont.

Seed coat infection

Viruses	Cucumber green mottle mosaic virus (cucumber) Tobacco mosaic virus (tomato mosaic virus) (tomato)
Fungi	*Cercospora kikuchii* (soybean) *Colletotrichum lindemuthianum* (bean) *Colletotrichum dematium* (soybean) *Drechslera (Helminthosporium) oryzae* (rice) *Verticillium albo-atrum* (sunflower)
Nematodes	*Ditylenchus dipsaci* (broad bean)

Seed surface contamination

Virus	Cucumber green mottle mosaic virus (cucumber) Tobacco mosaic virus (tomato mosaic virus) (tomato)
Fungi	*Drechslera graminea* (barley) *Drechslera teres* (barley) *Drechslera oryzae* (rice) *Peronospora manshurica* (soybean) *Tilletia caries* (wheat) *Tilletia foetida* (wheat)
Bacteria	*Curtobacterium flaccumfaciens* pv. *flaccumfaciens* (bean)

Concomitant contamination (plant tissues, debris, sclerotia, galls etc.)

Viruses	Cucumber green mottle mosaic virus (cucumber) Tobacco mosaic virus (tomato mosaic virus) (tomato)
Fungi	*Ascochyta rabiei* (chickpea) *Claviceps purpurea* (rye and other cereals) *Melanopsichium missouriense* (soybean) *Phoma medicaginis* var. *pinodella* (pea) *Sclerospora graminicola* (*Pennisetum typhoides*) *Sclerospora macrospora* (oat) *Sclerotinia sclerotiorum* (many crops) *Sclerotinia spermophila* (clover) *Typhula trifolii* (clover)
Bacteria	*Clavibacter michiganensis* subsp. *insidiosus* (alfalfa) *Pseudomonas syringae* pv. *phaseolicola* (bean)
Nematodes	*Anguina tritici* (wheat, rye) *Ditylenchus dipsaci* (red clover, lucerne) *Ditylenchus angustus* (rice) *Heterodera glycines* (soybean)

In general, it is advisable to discard defective seeds (abnormal, discoloured, small etc.), in order to reduce the percentage of possible seed-infection. Removal of any adherent plant or fruit debris, damaged seeds, weed and other seeds, inert materials etc., which in many cases carry inoculum, is necessary. Of course, the manual or mechanical removal of the abnormal or diseased-looking seeds is only the first step of pathogen elimination, and it does not provide complete elimination since many healthy looking seeds may also be infected.

Washing or surface sterilization. With surface adherent pathogens, it is practical to wash off the inoculum propa- gules from the seeds. Such are the cases of *Tilletia* spp. and *Urocystis tritici* spores in cereal seeds. The seeds have to be dried after the washing. Superficial infection or contamination of tomato seeds by tobacco mosaic virus (TMV), or cucumber seeds by cucumber green mottle mosaic virus (CGMMV), mainly from the fruit pulp, serving for seed transmission, can be removed by washing the seed thoroughly with water, or water containing detergents (such as Teepol 10%), followed by thorough rinsing and drying. The washing, however, does not have any effect on the internal infection of TMV and CGMMV.

Surface sterilization with 0.5% sodium hypochlorite containing 0.1% wetting agent for 10 minutes and rinsing thoroughly with water is a commonly used procedure.

Heat inactivation. Hot water treatments are used to eliminate various seed-borne pathogens. Immersing oat and barley seeds in hot water (52,8°C) for 5 minutes was used to eliminate *Ustilago avenae* (loose smut of oat) and *U. nuda* by J.L. Jensen as far back as in 1888 in Denmark. Seed-transmitted nematodes can be eliminated by hot water treatment (Brown and Kerry, 1987; Franklin and Siddiqui, 1972; Southey, 1972), as *A. tritici* in wheat (54°C for 10 min) and *Ditylenchus* spp. in various seeds (44-61°C for 10-15 min). Hot water treatments are used for the control of fungal and bacterial pathogens of vegetable seeds (Maude, 1983). Related to hot water treatment is the seed solarization applied by Beniwal *et al.* (1989) to reduce *Ascochyta lentis* inoculum from infected lentil seeds. Some farmers in the Middle East, even today, wash, soak, and expose seed to the sun. This is a method combining washing off and heat inactivation of the seed pathogen.

Hot water treatment has the advantage of being cheap and practical to apply in small seed lots, as in the case of valuable seed germplasm. Its disadvantage is that it may reduce the germinability of the seed, particularly in old seed samples (Maude, 1983).

Aerated steam treatment has also been used in the case of seeds. Its disadvantage is that it may reduce seed germination (Baker, 1969) like in hot water treatment. It is generally very effective with small seeds, but with larger seeds, such as peas, its efficiency is slightly less (Maude, 1966).

In the case of seed-transmitted viruses, heat treatment has been widely studied, but no effect has been found on viruses which are embryo-borne.

Heat treatment can completely eliminate TMV infection from tomato seeds and CGMMV infection from cucumber seeds. Seed transmission of cowpea banding mosaic virus is claimed to be considerably reduced in cowpea seed with hot water at 45°C for 40 minutes, or at 50°C for 20 minutes, or with hot air at 55°C for 50 minutes and at 65°C for 20 minutes. It is also reported to be completely prevented by dry heat treatment for 15 minutes at 65°C followed by 2,4, or 8 days at 30°C.

Heat treatment (usually 35-38°C for 5-15 weeks) is often combined with tissue culture, which is discussed later.

Gamma irradiation. Gamma irradiation has been partly successful in some cases. It partly disinfected barley seeds infected with barley stripe mosaic virus, when applied on seeds either dry or soaked in water for 18 hours. Certain inactivation of prunus necrotic ringspot virus (PNRV) and prune dwarf virus (PDV) from *Prunus* spp. seeds was obtained after gamma irradiation.

Aging. There are reports that infection in extra-embryonic tissues rapidly declines during storage, e.g. TMV in tomato seeds. During 5 years of storage of *Prunus* spp. seeds infected with PNRV or PDV, virus viability declined more rapidly than did the seed viability (Bos, 1977). Gradual virus inactivation was observed in seeds of eggplant varieties infected by a seed transmissible virus, total inactivation, after 7 months at room temperature, without any appreciable loss in seed germinability (Mayee, 1977). With viruses that infect the embryo, storage of seed does not seem to be of help in reducing seed transmission, since such viruses may survive for as long as the seeds remain viable (Bos, 1977).

In germplasm collections, some pathogens are inactivated during storage, at least down to some percentages. However, there are many examples where pathogens survive even under the recommended storage conditions. Some pathogens survive for more than 10 years. Kaiser and Hannan (1986) has reported recovery of *Ascochyta lentis* from lentil seed after storage of 30 years, and Pierce and Hungerford (1929) of bean common mosaic virus from bean seeds after 30 years.

Chemical treatments

Chemical treatments, especially when combined with soaking, may reach deeper tissues. To be effective against deep seated infections, chemicals must penetrate the tissues of seeds and kill the pathogen without causing phytotoxicity (Maude, 1983). Many chemicals used today, especially systemic fungicides, have this capacity, especially when applied after soaking seeds in aqueous solutions or suspensions of these chemicals.

Seed infection of peas by *Ascochyta pisi* was eliminated by soaking the seeds for 24 hours at room temperature in antibiotics (pimaricin and rimocidin) to achieve maximum penetration of seed (cotyledon) tissues. Aqueous suspensions of thiram at 30°C were used to soak pea seeds for 24 hours in order to achieve maximum eradication of *Ascochyta pisi*. The thiram soak treatment, as finally developed (0.2% aqueous thiram suspensions for 24 hours at 30°C, followed by seed drying) successfully controlled a large number of seed-borne diseases of vegetables and other crops (Maude, 1983).

Soaking in aqueous antibiotic solutions has been used to eliminate bacterial seed-transmitted pathogens. Streptomycin (200 ppm) for 24 hours at room temperature was successfully used against *Curtobacterium flaccumfaciens* pv. *betae* in red beet clusters, aureomycin, streptomycin and terramycin against *Xanthomonas campestris* pv. *campestris* in *Brassica* seeds. The phytotoxic effects of antibiotics on seeds could be neutralized by rinsing the antibiotic-treated seed with water, followed by immersion in 0.5% (w/v) sodium hypochlorite for 30 minutes. Eradication of *X. campestris* pv. *campestris* was achieved by increasing the soaking time to two hours (Maude, 1983).

Eradication of deep-seated fungal infections can also be obtained by seed surface application of systemic fungicides. This was initiated in 1966, when Von Schmeling and Kulka demonstrated the curative action of carboxin against *Ustilago nuda* in the embryos of barley seeds when applied to the seed surface (Maude, 1983). Systemic fungicides have many practical advantages. They are easy to apply as powders, slurries or in liquid form. They penetrate deeply into the seed tissues and they cause little or no phytotoxicity. Their main disadvantage is their selectivity for certain fungi, as compared to the broad spectrum activity of non-systemic fungicides. Those based on or closely related to benzimidazole structure, such as benomyl, carbendazim, thiophanate-methyl, and thiabendazole (TBZ), have been found very effective in eradicating pathogenic fungi from vegetable seeds. Applied as dusts or slurries, they controlled *Phoma lingam* (black leg) of brassica, *Ascochyta pisi* (leaf and pod spot) of peas, *Ascochyta fabae* (leaf spot) of *Vicia faba*, *Colletotrichum lindemuthianum* (anthracnose) of *Phaseolus vulgaris*, *Botrytis allii* (neck rot) of *Allium cepa*, *Fusarium oxysporum* f. sp. *ciceri* (wilt) of *Cicer arietinum* and *F. oxysporum* wilt of *Capsicum annuum* seeds (Maude, 1983).

In the case of the earlier mentioned chickpea blight caused by *Ascochyta rabiei* in the Regional Plant Introduction Station in Washington state, in laboratory tests with naturally infected seeds, the most effective seed treatments were achieved with benomyl and TBZ which reduced the incidence of seed-borne *A. rabiei* from 45% in the untreated control to 0 in the treated seeds. In laboratory tests, *A. rabiei* appeared to be eradicated from infected seeds treated with benomyl and TBZ, or combinations of these fungicides with captan. In greenhouse tests, using sterile potting medium, occasionally an infected seedling was detected in the benomyl and TBZ treatments, indicating that those chemicals do not completely eradicate *A. rabiei*. Sterile soil had to be used, or soil treatment with captan, to avoid pre-emergence damping-off by *Pythium ultimum* (Kaiser and Hannan, 1988).

Benzimidazole based systemic fungicides are not effective against greenhouse damping off of oomycetes. This disadvantage can be overcome by formulating these fungicides in mixtures with protectant fungicides such as thiram. A further limitation of the benzimidazole-based fungicides is their ineffectiveness against Dematiaceous

Hyphomycetes, such as *Alternaria*; alternative fungicides belonging to the dicarboximide group, particularly iprodione, have proved effective against such fungi in seeds (Maude, 1983).

Aqueous chemical soaking for 24 hours has the disadvantage that seeds become fully imbibed with water and so they require immediate drying, which is very difficult with commercial lots of large seeds such as peas and beans. To overcome such problems, the concept of applying pesticides in volatile organic solvents to seeds was developed. Such solvents, as acetone or dichloromethane, have been used for introducing DDT, cumarin and hormonal chemicals into the seeds, or systemic fungicides into the seed coats. Eradication has been obtained in some cases, and the seed germination is not impaired by organic solvents, with occasional exceptions (Maude, 1983). In the case of valuable seed germplasm, however, there does not seem to be a need for applying the chemicals with organic solvents since the seed quantities concerned are limited and manageable.

Biochemical (fermentation, pectolytic enzymes) or chemical (trisodium phosphate or hydrochloric acid) treatment, to remove external infection of seeds from fruit pulp remnants has been used against viruses such as tobacco mosaic virus (TMV) in tomato and cucumber green mottle mosaic virus (CGMMV) in cucumber. However, since the seed infection by these viruses is also present in the seed coat and even in the endosperm, the above treatments are combined with soaking to get complete seed disinfection. Polishing dry cucumber seeds and treating with detergents or chemicals reduced CGMMV infection but did not eradicate it. Tomato seed infection was greatly reduced by superficially cleaning them with 10% teepol, trisodium phosphate or hydrochloric acid, whereas the combined application of pectolytic enzymes and hydrochloric acid has also been suggested. In chemical treatments combined with soaking, TMV on or in the seed coats of tomato seeds was inactivated by treatment with 10% trisodium phosphate, or by extraction with concentrated hydrochloric acid (10 ml per 25 lb of fruit pulp, or 25 ml per 5 lb of fruits). The same virus was inactivated in tomato seeds and in their seed coats by soaking the seeds in 1% aqueous solution of sodium orthophosphate for 15 minutes, followed by soaking in 0.525% sodium hypochlorite for 30 minutes. Treatment with either chemical alone was insufficient (Bos, 1977). Soaking cowpea seeds infected by cowpea banding mosaic virus (a virus related to CMV) with malic hydrazide at 40-400 ppm for 90 minutes, 2-thiouracil at 500 and 700 ppm for 90 minutes, NAA at 40 ppm for 4 hours and Teepol (5-10%) for 4 hours completely inactivated the virus without affecting seed viability.

Sanitary inspection and selection and virus elimination schemes

Growing plants from the seeds of valuable seed germplasm and inspecting them during growth is an essential part of pathogen elimination procedure. Viruses (and viroids), in general, are the most difficult and in many cases impossible to eliminate from infected seeds due to their intimate association with and persistence in the infected cells. Nevertheless, seed germplasm sanitation from virus infection is possible by sanitary inspection and selection and, in general, easier than the vegetative germplasm resanitation. However, strangely enough, very little has been done, worldwide, on the production of virus-free true seeds and true seed certification, compared to the numerous certification programmes for vegetative germplasm.

Visual inspection, indicator plant test, ELISA and other serological techniques, such as ISEM, DNA- or RNA-hybridization etc., classical detection methods used today in the certification schemes of vegetative germplasm, can also be applied in seed sanitation programmes. Needless to say that the whole sanitation procedure, through field inspection and selection, should be carried out under conditions excluding infection from external sources (other than direct seed transmission), mainly by vectors. So, the seeds are grown in greenhouses or insect-proof screenhouses, or in isolated fields, in treated soil and under the proper chemical or other control measures.

There have been reported examples of successful virus-elimination schemes of valuable seed germplasm. Two of them are presented below.

Barley stripe mosaic virus was eliminated from the Composite Cross XXXI-B barley population at Bozeman, Montana, USA, by: 1) growing advanced generations in isolation (200 m) from other barley, 2) using the sensitive serological method of ELISA to detect BSMV in seed and leaves, 3) testing all plants of the most advanced generation (Sib 18, F4 and Sib 18, F5) and a large sample of the seed by ELISA, and 4) destroying all infected plants and contiguous, apparently healthy plants within the same planting block. BSMV was detected in 1966 but by 1987 the population was clean of BSMV, using the above procedure (Carroll *et al.*, 1990).

Systematic was also the work of cleaning the USA pea collections from pea seed-borne mosaic virus. One hundred accessions of *Pisum* germplasm known to contain PSbMV were tested experimentally as models of germplasm reclamation. Half of them were planted in Corvallis, Oregon, and half in Pullman, Washington, USA, 75 seeds each. Most of those accessions showed identifiable symptoms of PSbMV in 1-7 seedlings which were eliminated by roguing and destroyed by autoclaving. Twenty to 30 potential mother plants were selected per accession and they were subjected to PSbMV detection assays by ELISA. All plant sets with latent PSbMV infection (these occurred in 17 of 100 accessions) were either discarded or retested per individual plant, in order to save those that were virus-free. Candidate germplasm mother plants were then grown to maturity, further examined for virus-like symptoms and assayed by ELISA a second time just before terminal leaves became senescent. Only 3 out of 100 accessions were found to contain PSbMV-infected plants at this stage. Seeds harvested from the selected plants were planted for an identical second generation series of tests to determine the success of the first generation. No virus-infected plants were detected in these progenies indicating that ELISA was sufficiently sensitive to detect all PSbMV infected pea plants in a single generation of testing (Hampton, 1983).

Tissue culture

Tissue culture, especially shoot tip culture, has been established as an efficient method for raising pathogen-free plants. Primarily it is directed against viruses, viroids, MLO's and RLO's, but also against some systemic infections of bacteria and fungi. Today it is the essential step of certification schemes for vegetatively propagated material, especially from producing the elite or nuclear stocks.

The method is scientifically based on the existing possibility of obtaining pathogen-free progeny (up to a certain percentage) from a systemically infected plant, by culturing *in vitro* abcised meristems composed of the apical dome and 2 leaf primordia. It is often complemented by thermotherapy (35-38°C for 5 to 15 weeks, prior or during the tissue culture) against spherical viruses, for enlarging the virus-free meristematic zone (Quak, 1977). In recent years it is often complemented with chemotherapy, using nucleic acid base analogues like virazole (ribavirin) or antimetabolites like vidarabine in the nutrient medium for suppressing virus multiplication.

Tissue culture, besides pathogen eradication, has also other practical uses such as quick multiplication of valuable breeding material, screening for genetically controlled defects, facilitating importation and exportation of plant material, and theoretical uses as synchronization of virus multiplication etc. have been applied in plant species propagated by true seed (Bhojwani and Razdan, 1983; Conger, 1981; Dale, 1977; Kartha *et al.*, 1974; White *et al.,* 1977) and the method has been tried for the elimination of infection in germplasm of true seed (Barnett *et al.*, 1975; Walkey and Cooper, 1975; Walkey *et al.*, 1974).

Testing for infection

Precise detection, identification and characterization of pathogens is an essential prerequisite of the pathogen elimination from valuable seed germplasm. As a rule, it is applied at the beginning and at the end of the elimination procedure, and usually, also during the procedure. It is done at the beginning in order to learn if the seed lot is carrying seed-borne pathogens, and if so, which ones, and to what extent the seed infection is present, so that the proper elimination method can be applied. It is done at the end in order to make sure that the elimination procedure succeeded and to what extent.

For details on seed health testing tehcniques, the reader is suggested to consult the section on this topic included in these proceedings. All tests should be carried out under meticulously controlled conditions, securing exclusion of infection by the target pathogen(s) or other pathogens from external sources. In the case of growing plants from the seeds, the plants should be grown in well controlled greenhouses, screenhouses, or isolated fields, and they should be provided with all necessary preventive measures (soil treatment, protective sprays etc.).

References

Agarwal, V.K. and Sinclair, J.B. 1987. *Principles of Seed Pathology*. Vol I. CRC Press Inc., Boca Raton, Florida, U.S.A. 176 pp.

Agarwal, V.K. and Sinclair, J.B. 1987. *Principles of Seed Pathology*. Vol. II. CRC Press Inc., Boca Raton, Florida, U.S.A. 168 pp.

Agarwal, V.K., Nene, Y.L., Beniwal, S.P.S. and Verma, H.S. 1979. Transmission of bean common mosaic virus through urdbean seeds. *Seed Science and Technology* **7**: 103-108.

Baker, K.F. 1969. Aerated seed treatment of seed for disease control. *Horticultural Research* **9**: 59.

Barnett, O.W., Gibson, P.B. and Seo, A. 1975. A comparison of heat treatment, cold treatment, and meristem tip culture for obtaining virus-free plants of *Trifolium repens. Plant Disease Reporter* **59**: 834-836.

Beniwal, S.P.S., Chanbey, S.N. and Bharathan, N. 1980. Presence of urdbean leaf crinkle virus in seeds of mungbean germplasm. *Indian Phytopathology* **33**: 360-361.

Beniwal, S.P.S., Seid, A. and Tadesse, N. 1989. Effect of sun drying of lentil seeds on the control of *Ascochyta lentis. Lens Newsletter* **16**(2): 27-28.

Bhojwani, S.S. and Razdan, M.K. 1983. *Plant Tissue Culture. Theory and Practice*. Elsevier. 502 pp.

Bos, L. 1977. Seed-borne pathogens. pp. 39-69. In: *Plant Health and Quarantine in International Transfer of Genetic Resources* (eds. W.B. Hewitt and L. Chiarappa). CRC Press Inc., Cleveland Ohio.

Brown, R.H. and Kerry, B.R. 1987. *Principles and Practice of Nematode Control in Crops*. Academic Press. 447 pp.

Carrol, T.W., Hockett, E.A., and Zaske, S.K. 1990. Elimination of seed-borne barley stripe mosaic virus (BSMV) from barley. *Seed Science & Technology* **18**: 405-414.

Chiarappa, L. and Gambogi, P. 1986. Seed pathology and food production. *FAO Bulletin* **34**(4): 166-185.

C.M.I./A.A.B. Descriptions of Pathogenic Fungi and Bacteria

C.M.I./A.A.B. Descriptions of Plant Viruses.

Conger 1981. *Cloning Agricultural Plants via In Vitro Techniques*. CRC Press Inc., Boca Raton, Florida: 1-273.

Dale, P.J. 1977. The elimination of ryegrass mosaic virus from *Lobium multiflorum* by meristem tip culture. *Annals of Applied Biology* **85**: 93-96.

Davis, R.F., Weber, Z., Pospieszny, H. and Silbernagel, M. 1981. Seed-borne cucumber mosaic virus in selected *Phaseolus vulgaris* germplasm and breeding lines in Idaho, Washington and Oregon. *Plant Disease* **65**: 492-494.

Diekmann, M. 1990. Seed health testing and treatment of germplasm at the International Center for Agricultural Research in the Dry Areas (ICARDA). *Seed Science & Technology* **16**: 405-417.

Drijfhourt, E. 1978. Genetic interaction between *Phaseolus vulgaris* and bean common mosaic virus, with implications for strain identification and breeding for resistance. Agricultural Research Report (Versl. landbouwk. Onderz.) (872, ISBN 902200671 9, vii).

Franklin, M.T. and Siddiqui, M.K. 1972. *Aphelenchoides besseyi*. CIH Descriptions of Plant-parasitic Nematodes. Set. 1, No.4, 3pp.

Goodman, R.M. and Oard, J.H. 1980. Seed transmission and yield losses in tropical soybeans infected by soybean mosaic virus. *Plant Disease* **64**: 913-914.

Grogan, R.G. 1983. Lettuce mosaic virus control by use of virus-indexed seed. *Seed Science & Technology* **11**: 1043-1049.

Hampton, R.O. 1982. Incidence of the lentil strain of pea seed-borne mosaic virus as a contaminantfo *Lens culinaris* plasm. *Phytopathology* **72**: 695-698.

Hampton, R.O. 1983. Seed-borne viruses in crop germplasm resources: disease dissemination risks and germplasm reclamation technology. *Seed Science & Technology* **11**: 535-546.

Hampton, R.O. and Braverman, S.W. 1979. Occurrence of pea seedborne mosaic virus in North American pea breeding lines, and new virus-immune germplasm in the Plant Introduction Collection of *Pisum sativum. Plant Disease Reporter* **63**: 95-99.

Hockett, E.A., Carroll, T.W. and Zaske, S.K. 1988. Registration of barley composite cross XXXI-A and B. *Crop Science* **28**: 722.

Kaiser, W.J. and Hannan, R.M. 1982. *Ascochyta lentis*: Incidence and transmission in imported lentil seed. *Phytopathology* **72**: 944. (Abstract).

Kaiser, W.J. and Hannan, R.M. 1986. Incidence of seed-borne *Ascochyta lentis* in lentil germplasm. *Phytopathology* **76**: 355-360.

Kaiser, W.J. and Hannan, R.M. 1988. Seed transmission of *Ascochyta rabiei* in chickpea and its control by seed-treatment fungicides. *Seed Science & Technology* **16**: 625-637.

Kartha, K.K., Gamborg, O.L. and Constabel, F. 1974. Regeneration of pea plants from shoot apical meristems. *Zeitzchift fur Planzenphysiologie* **72**: 172-176.

Maude, R.B. 1966. Testing steam/air mixtures for control of *Ascochyta pisi* and *Mycosphaerella pinodes* on pea seed. *Plant Pathology* **15**: 187-189.

Maude, R.B. 1983. Eradicative seed treatments. *Seed Science & Technology* **11**: 907-920.

Mayee, C.D., 1977. Storage of seed for pragmatic control of a virus causing mosaic disease of brinjal (eggplant). *Seed Science & Technology* **5**: 555-558.

Neergaard, P. 1979. *Seed Pathology*. Vol. I and II. The Macmillan Press Ltd., London and Basingstoke. 1191 pp.

Pierce, W.H. and Hungerford, C.W. 1929. A note on the longevity of the bean mosaic virus. *Phytopathology* **19**: 605.

Quak, F. 1977. *In vitro* **13**: 194.

Smith, I.M., Dunez, J., Lelliott, R.A., Phillips, D.H. and Archer, S.A. (eds.) 1988. In: *European Handbook of Plant Diseases*. Blackwell Scientific Publications. 583 pp.

Southey, J.F. 1972. *Anguina tritici. C.H.I. Descriptions of Plant-parasitic Nematodes*. Set **1** No.13, Commonwealth Institute of Parasitology. St. Albans, Herts., England. 4pp.

Walkey, D.G.A. and Cooper, V.C. 1975. Effect of temperature on virus eradication and growth of infected tissue cultures. *Annals of Applied Biology* **80**: 185-190.

Walkey, D.G.A., Cooper, V.C. and Crisp, P. 1974. The production of virus-free cauliflowers by tissue culture. *Journal of Horticultural Science* **49**: 273-275.

White, J.L., Wu, F.S. and Murakishi, H.H., 1977. The effect of low temperature treatment of tobacco and soybean callus cultures on rates of tobacco and southern bean mosaic virus synthesis. *Phytopathology* **67**: 60-63.

Geophytopathology of Some Seed-borne Diseases

MARLENE DIEKMANN

International Center for Agricultural Research in the Dry Areas (ICARDA), P.O.Box 5466, Aleppo, Syria

Introduction

Forecasting of outbreaks of plant diseases is based on the nature of pathogens and meteorological data of a given area. According to Bourke (1970) four criteria are necessary to develop a forecasting system and they are:

- economic importance of the disease
- appreciable impact of weather factors on disease variability
- availability of information on weather dependence and
- availability of control measures

Successful forecasting systems were developed for late blight of potatoes (Waggoner, 1968), apple scab (Kranz *et al.*, 1973), barley powdery mildew (Aust *et al.*, 1983) and for cereal diseases (Polley and Clarkson, 1978). Ideally they enable the farmer to decide whether a treatment is economic and what is the optimal time of application.

In these forecasting systems weather data are used rather than climatological data. But the use of climatological data in combination with data on plant pathogens and their distribution led to "geophytopathology". Weltzien (1967) created this term, and suggested to consider distribution of the host plant, frequency of disease occurrence, its intensity, and the extent of damage. He further proposed differentiation in: (1) "areas of main damage", where epidemics with considerable economic losses occur frequently, (2) "areas of marginal damage", where epidemics with considerable economic losses occur occasionally and (3) "areas of sporadic attack", where the disease is reported, but does not usually cause significant damage.

Maps following these criteria were subsequently published by Drandarevski (1969) for *Erysiphe betae* and Bleiholder and Weltzien (1972) for *Cercospora beticola.*

Maps of plant diseases based only on reports on the occurrence of a disease started much earlier, such as the CMI Distribution Maps of Plants Diseases, which have been issued by the Commonwealth Agricultural Bureaux since 1942. This loose leaf collection contained in 1989 maps on 616 different pathogens. Each year about 10 to 15 new maps are added, and about 30-40 maps are updated. The information included in these maps comes mostly from articles summarized in the Review of Plant Pathology (formerly Review of Applied Mycology). However, a country not listed is not necessarily free from a particular disease; or as the saying goes: absence of evidence is not evidence of absence. In Syria, for example, faba beans are greatly affected by *Uromyces fabae*, *Botrytis fabae* and *Ascochyta fabae,* however, the first publication accessible through the common literature search systems was in 1979 by Hawtin and Stewart. This is the reason why Syria was not included in the respective maps (Commonwealth Agricultural Bureaux 1977a, 1981, 1976, Fig. 1, 2, and 3). The second edition of the *Ascochyta* map (Commonwealth Agricultural Bureaux 1989, Fig. 4) has included Syria.

Collection of data on disease requires frequent field surveys by experienced personnel so that early infection can be noticed. Diseases may be also measured indirectly, e.g. by monitoring the release of spores. Young *et al.* (1978) gave an overview of the various methods ranging from slides coated with sticky material to sophisticated samplers. Leppik (1964) complemented the information available in the current CMI Distribution Map for *Plasmopara halstedii* with the results of his seed health tests on sunflower seeds originating from various parts of the world. This more than doubled the number of host countries. In the maps published subsequently, this information has been incorporated (Commonwealth Agricultural Bureaux, 1977b).

Unfortunately, a lot of valuable data are not published at all, or published only in annual reports and other publications which are not easily accessible to the scientific community. There appears to be a tendency that scientists spending a great part of their time in the field do not find time to report their findings in widely circulated publications.

Most researches are carried out on diseases that cause considerable economic losses, whereas it is difficult to find reports in literature on those occurring only sporadically. Yet the latter may be the most interesting because it is subject to changes more often than the former. It is in this group that we find diseases introduced into uninfected areas, and they may gain importance rapidly. This depends, among other factors, on the weather conditions and the agricultural and legislative practices. Once diseases get established in new areas their eradication may be very expensive. Examples are the eradication campaigns carried out against citrus canker in Florida at a cost of 25 million US$ from September 1984 to July 1986 (Schoulties *et al.*, 1987), and the Mediterranean fruit fly in California at 100 million US$ from 1980 to 1982 (Johnson and Schall, 1989). An early detection and quick action can reduce the cost of eradication considerably, e.g. only 4.4 million US$ were spent in the eradication programme against the Mediterranean fruit fly in Florida during 1981 to 1985 (Johnson and Schall, 1989).

One of the major applications of geophytopathology is the assessment of the risk that a disease might occur in an area where it had not been reported hitherto. This aspect is important when new crops are introduced through the international seed exchange (Weltzien, 1988).

Sufficient time has elapsed since prognoses have been published and it is time that they should be examined now. Some examples are:

Powdery mildew of sugar beet (*Erysiphe betae*)

Drandarevski (1969) found that although there were no reports in the literature on mildew on sugar beets in Finland, Bulgaria, Yugoslavia, Albania, Greece, the Iberian Peninsula, North Africa, Southeast Asia, South America and Australia, the climatic conditions were such that epidemics could be expected in some of them. Moreover, for the United States, where the disease occurred but was not reported to be of importance, an increased importance in future was suspected. A literature review covering the years 1968 to 1988 produced reports on the occurrence of the pathogen in Bulgaria (V"rbanov, 1978), Egypt (El-Kazzaz *et al.*, 1977), and Portugal (Lucas *et al.*, 1979). Amano (1986) listed the following additional host countries: Yugoslavia, Spain, Morocco, and Japan, unfortunately all without direct reference to the original source.

Furthermore, an epidemic of sugar beet powdery mildew occurred in 1974 in USA (Ruppel *et al.*, 1975), with severe attacks in California, Utah, Idaho, Oregon, Washington, Northwest Kansas, and Nebraska and moderate attacks in most of the Great Plains. Drandarevski's map projected areas of significant economic importance in California, Utah, and Idaho, and areas of occasional significant importance for the other Western beet producing states. In 1975, the pathogen was observed for the first time in Michigan, without causing economic losses (Schneider and Hogaboam, 1977). Drandarevski (1969) had projected Michigan as an area where the pathogen could occur without economic importance. Also in 1975, powdery mildew on sugar beet was reported for the first time in Canada (Harper and Bergen, 1976). The authors supposed that there was little danger as the mycelium of the fungus could not overwinter in Canada. No reports on the occurrence of powdery mildew on sugar beets have been published in Canada during 1976 and 1988.

Tobacco blue mould (*Peronospora tabacina*)

Miller had concluded already in 1950 from his studies on forecasting the incidence of tobacco blue mould caused by *Peronospora tabacina*, in the United States that the pathogen would cause a serious disease if it is imported to Europe (Miller, 1969). His prediction was proven accurate by the blue mould epidemics which started from an introduction from Australia to England in 1958 and spread from there in four years all over Europe (Weltzien, 1981).

Powdery mildew and leaf spot of suagr beet

A "negative prognosis" was made by Weltzien (1978) by combining the data of Drandarevski (1969) and Bleiholder and Weltzien (1972) with the distribution of sugar beets (Fig. 5). He concluded that the northern coastal areas of Europe, the East coast of the USA, and an area in Northeast Argentina were not likely to suffer from either *Cercospora beticola* or *Erysiphe betae*. For *E. betae* no reports from these areas could be found, and for leaf spot the search yielded only one report mentioning a *Cercospora beticola* isolate from Maryland (Whitney and Lewellen, 1976). It can therefore be assumed that even though the pathogens might occur in the areas mentioned by Weltzien (1978) they are not considered important enough to warrant research to be conducted and published.

Mathematic models are used frequently as a tool in the forecasting of plant disease epidemics. Kranz and Royle (1978) gave a critical perspective of modelling in epidemiology. So far, however, there has been little use of mathematical models in geophytopathology. Some recent studies on the bacterial leaf blight of rice (BLB) caused by *Xanthomonas campestris* pv. *oryzae* and Ascochyta blight of chickpea caused by *Ascochyta rabiei* are presented below for discussion (Diekmann, 1991; Diekmann and Bogyo, 1991).

Monthly data for mean daily maximum and mean daily minimum temperature, mean precipitation and mean number of days with precipitation, and mean wind speed from standard meteorological stations (Müller, 1982) were analyzed. Fig. 6 and 7 indicate the location of the stations as well as the reported distribution area of Ascochyta blight and bacterial leaf blight, respectively. Data for the month of planting and the subsequent three (Ascochyta blight) and four (BLB) months were considered. A multiple linear regression analysis was made first, and regression coefficients of 0.88 (Ascochyta blight) and 0.68 (BLB) were found (Fig. 8 and 9). Some variables that showed a high correlation with the disease rating were plotted against the ratings as well as against the predicted values (Fig. 10 and 11). In order to identify the variables which contribute most to a discrimination between the two groups, a step-wise discriminant analysis (SDA) was conducted using BMDP (Jennrich and Sampson, 1985). This method permits the study of the differences between two or more groups with respect to a number of variables simultaneously. Discriminant functions combine group characteristics so that the group (here: disease-risk and non-disease-risk) which the case (here: location) most resembles can be identified. Histograms of the canonical variable are shown in Fig. 12 and 13. The "misclassifications" indicate that based on the climate data there is no risk for BLB at Wuhan, China and no risk for Ascochyta blight at Guaymas, Mexico. These areas can probably be considered "areas of sporadic attack" in the sense of Weltzien (1967). The cases where a potential disease risk was indicated, based on the climate data, are particularly interesting. For BLB these are Montgomery, Alabama and Charleston, South Carolina. The disease was reported for the first time in Texas in 1987 and Louisiana in 1988 (Jones *et al.*, 1989) and it could be a matter of time for the pathogen to spread to a so far disease-free location. Ascochyta blight was introduced with infected seed to Eastern Washington and Northern Idaho (Kaiser and Muehlbauer, 1984) in the early eighties, and according to the climate analysis the area of Fresno, California also has a suitable climate for the disease.

Prediction of disease risk at other locations

Ascochyta blight of chickpea (*Ascochyta rabiei*)

From the classification functions the following discriminant function was computed:

$$y = -1.11 - 0.22\,x_1 - 0.05\,x_2 - 0.32\,x_3 + 0.91\,x_4 - 0.51\,x_5 + 1.15\,x_6$$

x_1 = mean daily temperature in month 1 of the vegetation
x_2 = mean precipitation in month 2 of the vegetation
x_3 = average precipitation per rainy day in month 1 of the vegetation
x_4 = average precipitation per rainy day in month 2 of the vegetation
x_5 = mean number of rainy days in month 1 of the vegetation
x_6 = mean number of rainy days in month 2 of the vegetation

The disease risk can be estimated for any location if the above parameters are known. If the computed score is > 0, the location is to be classified as "disease risk", if it is < 0, it is "no disease risk". Table 1 gives some examples for the application of this method.

Table 1. Examples for the prediction of Ascochyta blight risk in different locations or planting time.

Area	Month of planting	x_1	x_2	x_3	x_4	x_5	x_6	score	disease risk
Plovdiv, Bulgaria	April	12.2	55	6.1	6.1	7	9	3.84	+
Kenyan highlands	June	15.7	17	5.8	3.4	5	5	-0.98	-
Bogota, Colombia	April	13.7	105	5.3	5.0	19	21	7.94	+
Neustadt, Germany	April	10.0	50	3.4	3.8	14	13	4.37	+
Aleppo, Syria	March	10.9	28	5.4	7.0	7	4	0.76	+
Aleppo, Syria	April	16.4	8	7.0	4.0	4	2	-3.46	-

Bacterial leaf blight of rice (*Xanthomonas campestris* pv. *oryzae*)

For the classification of locations into "disease risk" or "no disease risk" a discriminant function was calculated from the classification functions:

$$y = -14.16 - 0.29\,x_1 + 1.19\,x_2 - 0.01\,x_3$$

x_1 = mean daily maximum temperature in month 1 of the vegetation
x_2 = mean daily minimum temperature in month 2 of the vegetation
x_3 = mean precipitation in month 3 of the vegetation

If the computed score for a specific location is negative, the location is classified as "no disease risk", if it is positive, the location is classified as "disease risk". Table 2 gives some examples for this application of the method.

Table 2. Examples for the prediction of BLB risk in different locations or planting time

Area	Month of planting	x_1	x_2	x_3	score	disease risk
Nagasaki, Japan	June	25.6	22.8	189	3.66	+
Hanoi, Vietnam	May	32.2	25.6	76	6.21	+
Pusan, South Korea	June	23.9	21.7	156	3.17	+
Makurdi, Nigeria	July	29.4	22.2	279	0.94	+
Eala, Zaire	June	30.6	17.8	178	-3.63	-
Venice, Italy	May	21.0	17.1	52	-0.42	-
Samsun, Turkey	July	26.1	18.3	61	-0.56	-
Samsun, Turkey	June	23.3	18.7	33	1.00	+

Those areas for which a disease score of close to 0 was computed, i.e. those areas for which a good discrimination could not be found, could probably be considered "areas of sporadic attack" in the sense of Weltzien (1967).

This method could also be applied to other host/pathogen combinations, as well as to pests.

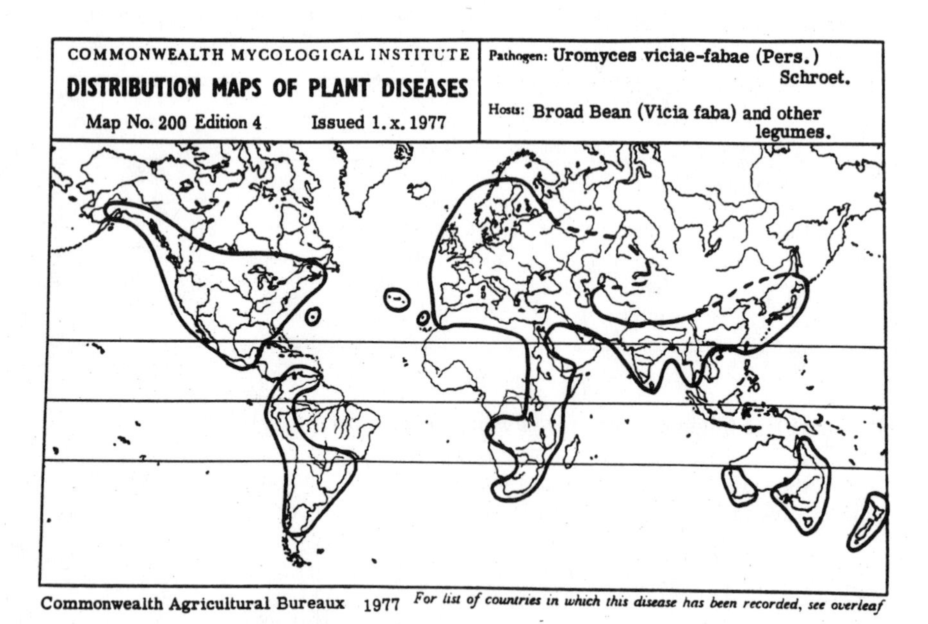

Fig. 1. CMI Distribution Map: *Uromyces viciae–fabae*

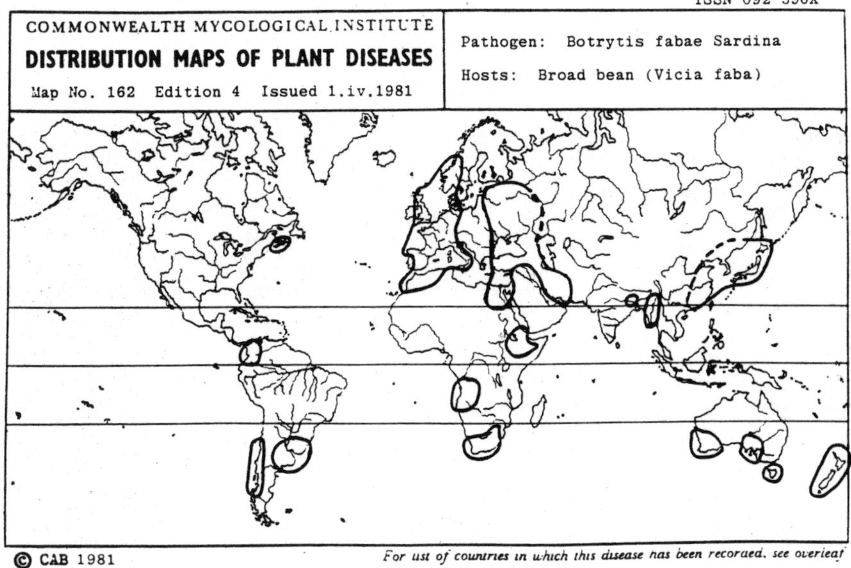

Fig. 2. CMI Distribution Map: *Botrytis fabae*

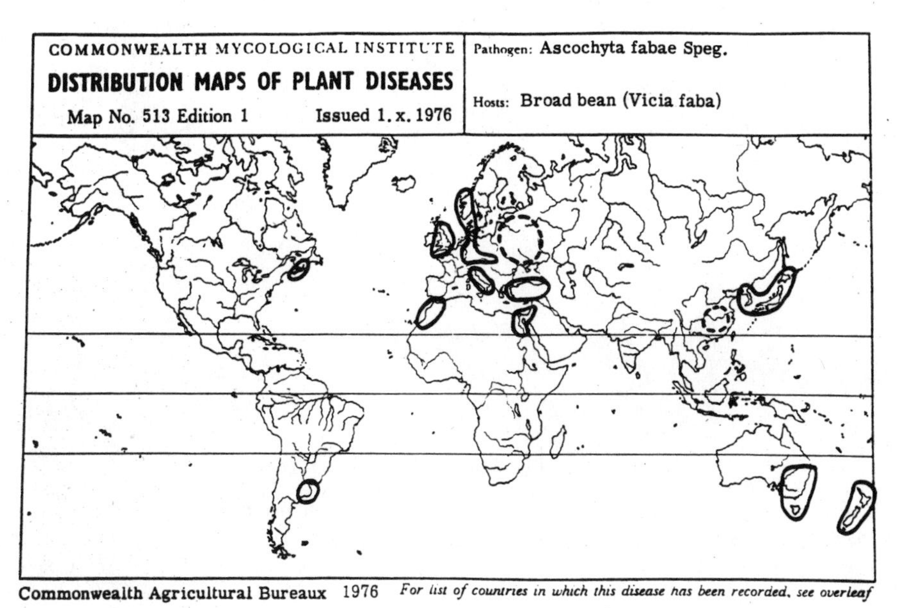

Fig. 3. CMI Distribution Map: *Ascochyta fabae*, edition 1

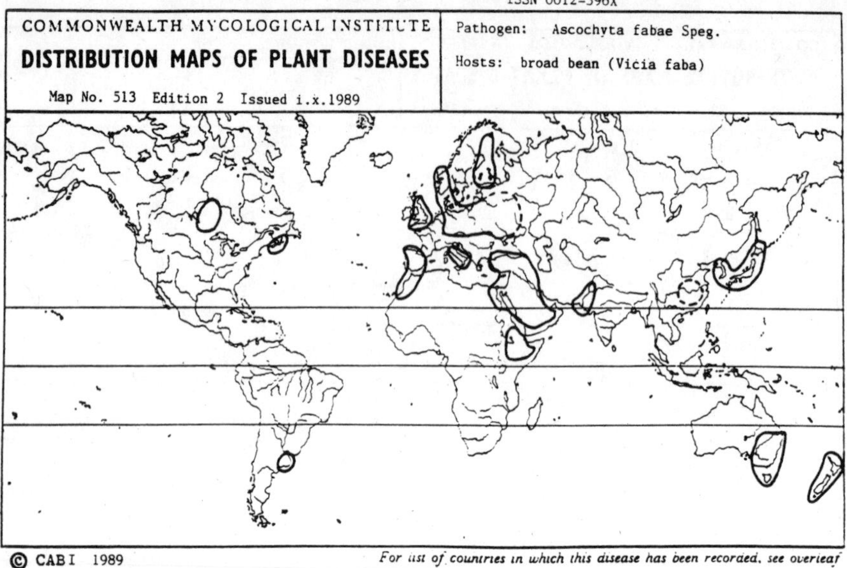

Fig. 4. CMI Distribution Map: *Ascochyta fabae*, edition 2

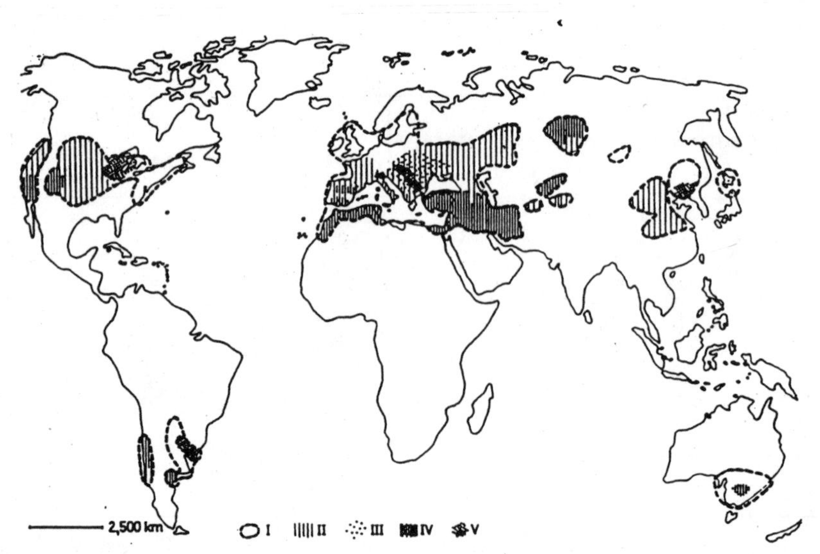

Fig. 5. World map of sugar-beet-growing areas with the three zones of disease intensity for leaf spot (*Cercospora beticola*) and powdery mildew (*Erysiphe betae*). I. Areas of sporadic attack for powdery mildew and leaf spot (Zone III). II. Areas of occasional powdery mildew epidemics (Zone II). III. Areas of occasional leaf spot epidemics (Zone II). IV. Areas of regular powdery mildew epidemics (Zone I). V. Areas of regular leaf spot epidemics (Zone I). (Weltzien 1981)

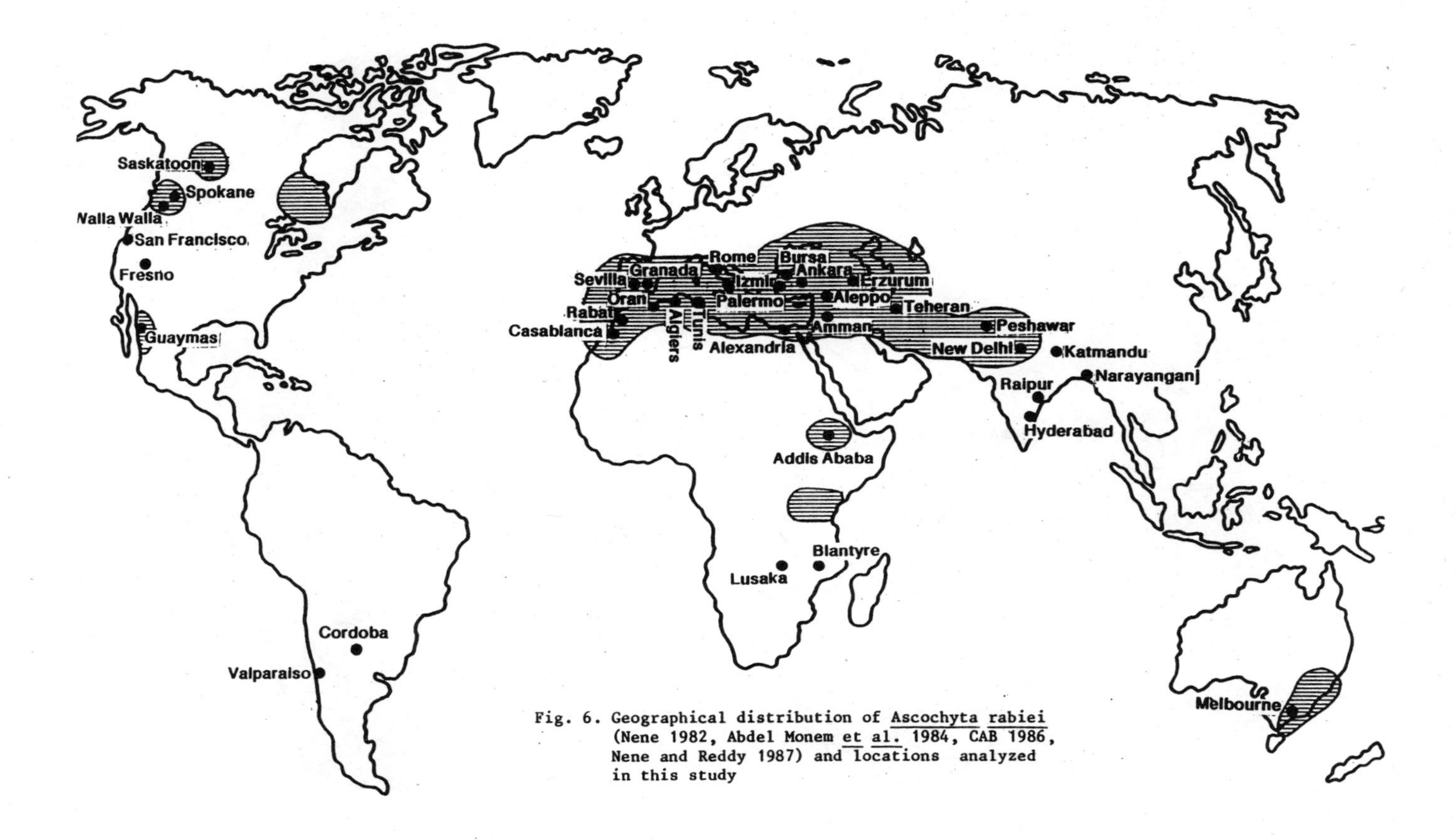

Fig. 6. Geographical distribution of *Ascochyta rabiei* (Nene 1982, Abdel Monem *et al.* 1984, CAB 1986, Nene and Reddy 1987) and locations analyzed in this study

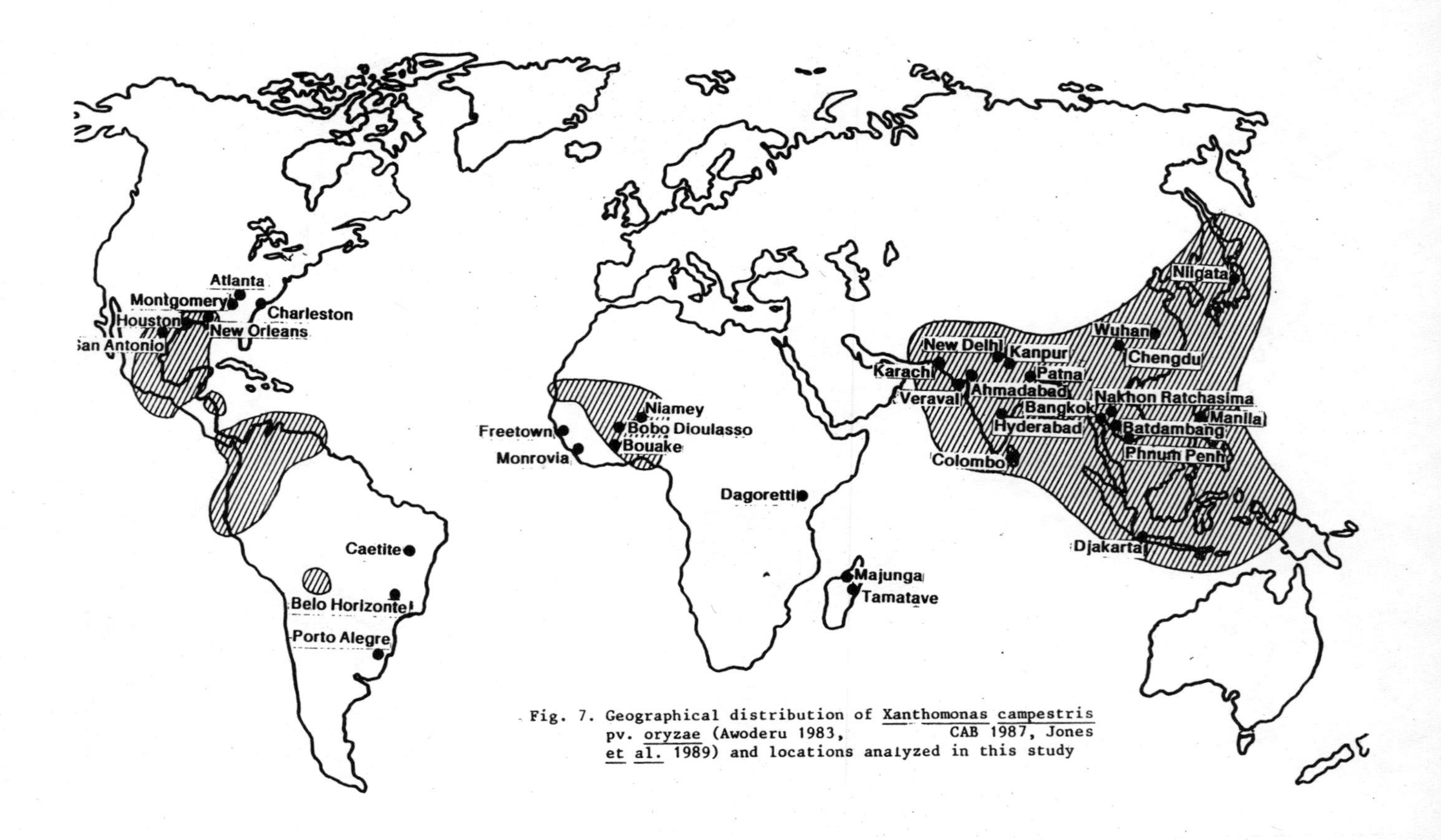

Fig. 7. Geographical distribution of Xanthomonas campestris pv. oryzae (Awoderu 1983, CAB 1987, Jones et al. 1989) and locations analyzed in this study

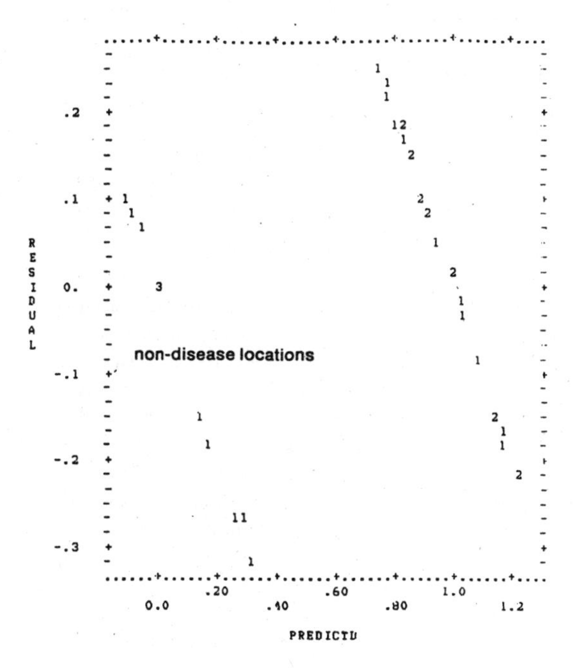

Fig. 8 .*Ascochyta* blight of chickpea; multiple linear regression between disease / non-disease location and climatic parameters.
The residuals ($y_i - \hat{y}_i$) are plotted against the predicted values ($\hat{y}_i$).
multiple r^2 = 0.8805

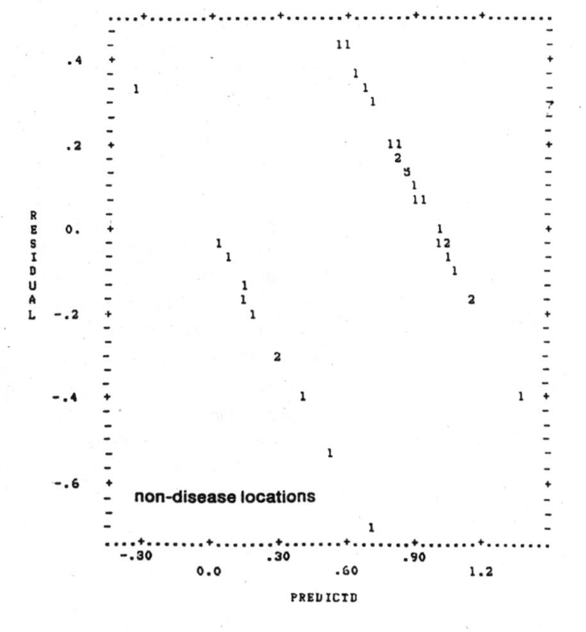

Fig. 9 . Bacterial leaf blight of rice; multiple linear regression between disease / non-disease location and climatic parameters.
The residuals ($y_i - \hat{y}_i$) are plotted against the predicted values ($\hat{y}_i$).
multiple r^2 = 0.6790

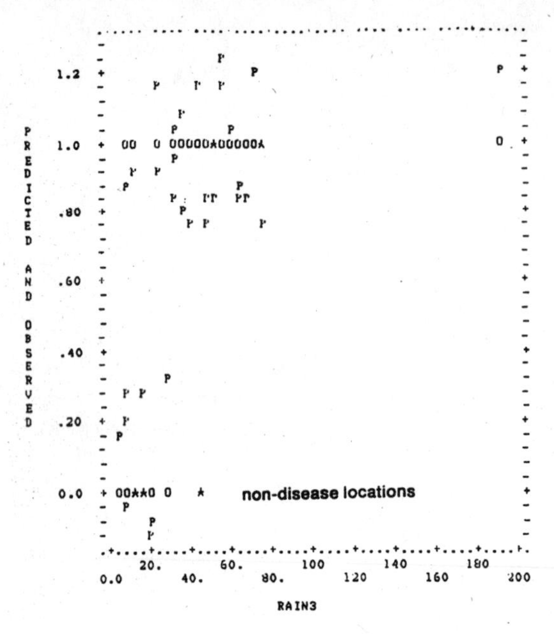

Fig.10 ***Ascochyta*** **blight of chickpea; disease values 0 (non-disease) and 1 (disease) as well as predicted values are plotted against the variable "rain in month 3 of the vegetation".**
O = observed values
P = predicted values
*** = overlap between observed and predicted values**

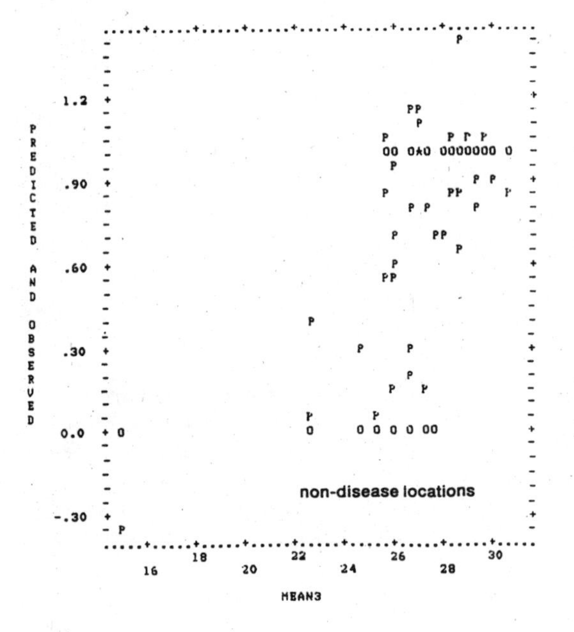

Fig.11 Bacterial leaf blight of rice; disease values 0 (non-disease) and 1 (disease) as well as predicted values are plotted against the variable "mean temperature in month 3 of the vegetation".
O = observed values
P = predicted values
*** = overlap between observed and predicted values**

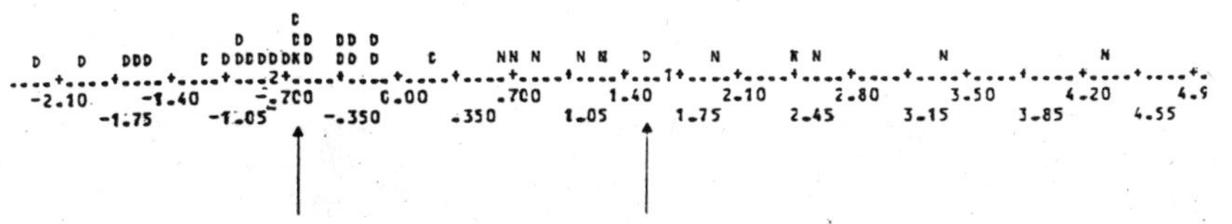

Fig. 12. *Ascochyta* blight of chickpea; histogram of the canonical variable,
N = non-disease risk, D = disease risk

misclassifications are marked
Guaymas, Mexico (can. v. = 1.52) falls into the "non-disease" group
Fresno, U.S.A. (can. v. = -0.64) falls into the "disease" group

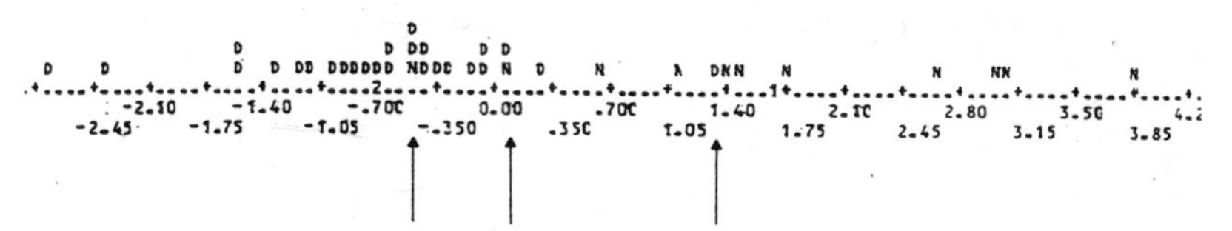

Fig. 13. Bacterial leaf blight of rice; histogram of the canonical variable,
N = non-disease risk, D = disease risk

misclassifications are marked
Wuhan, China (can. v. = 1.34) falls into the "non-disease" group
Montgomery and Charleston, U.S.A. (can. v. = 0.08 and -0.52) fall into the "disease" group

References

Abdelmonem, A.M., Yehia, A.H. and El Wakil, A.A. 1984. *Ascochyta rabiei*, a new seed-borne pathogen of chickpea in South Tahreer, Egypt. *Egyptian Journal of Phytopathology* **16**: 1-10.

Amano, K. 1986. *Host Range and Geographical Distribution of the Powdery Mildew Fungi*. Japan Scientific Societies Press, Tokyo.

Aust, H.-J., Hau B. and Kranz, J. 1983. Epigram - a simulator of barley powdery mildew. *Z. Pflanzenkr. Pflanzenschutz* 90: 244-250.

Awoderu, V.A. 1983. Technical Service Support: Pathology. Paper presented at the Annual Rice Review Meeting WARDA/82/ARR-12C, Monrovia.

Bleiholder, H. and Weltzien, H.C. 1972. Beiträge zur Epidemiologie von *Cercospora beticola* Sacc. an Zuckerrübe. III. Geopathologische Untersuchungen. *Phytopath. Z.* **73:** 93-114.

Bourke, P.M.A. 1970. Use of weather information in the prediction of plant disease epiphytotics. *Ann. Rev. Phytopathol.* **8:** 345-370.

Commonwealth Agricultural Bureaux. 1976. *Ascochyta fabae* Speg. *CMI Distribution Maps of Plant Diseases*. Map No. 513.

Commonwealth Agricultural Bureaux 1977a. *Uromyces viciae-fabae* (Pers.) Schroet. *CMI Distribution Maps of Plant Diseases*. Map No. 200. 4th ed.

Commonwealth Agricultural Bureaux 1977b. *Plasmopara halstedi* (Farl.) Berl. & de Toni. *CMI Distribution Maps of Plant Diseases*. Map No. 286. 4th ed.

Commonwealth Agricultural Bureaux 1981. *Botrytis fabae* Sardina. *CMI Distribution Maps of Plant Diseases*. Map No. 162. 4th ed.

Commonwealth Agricultural Bureaux 1986. *Ascochyta rabiei* (Pass.) Labrousse. *CMI Distribution Maps of Plant Diseases*. Map No. 151. 4th ed.

Commonwealth Agricultural Bureaux 1987. *Xanthomonas campestris* pv. *oryzae* (Ishiyama) Dye. *CMI Distribution Maps of Plant Diseases*. Map No. 304. 5th ed.

Commonwealth Agricultural Bureaux 1989. *Ascochyta fabae* Speg. *CMI Distribution Maps of Plant Diseases*. Map No. 513. 2nd ed.

Diekmann, M. 1991. Occurrence of Ascochyta blight on chickpea depending on climatic conditions. Submitted for publication to Plant Disease.

Diekmann, M. and Bogyo, T.P. 1991. Distribution of bacterial leaf blight of rice (*Xanthomonas campestris* pv. *oryzae* (Ishiyama) Dye) depending on climatological factors. Submitted for publication to Journal of Plant Diseases and Protection.

Drandarevski, C.A. 1969. Untersuchungen über den echten Rübenmehltau *Erysiphe betae* (Vanha) Weltzien. *Phytopath. Z.* **65**: 201-218.

El-Kazzaz, M.K., Mehiar, F.F. and El-Khadem, M.M. 1977. Powdery mildew on sugar beet a new record for Egypt. *Egyptian Journal of Phytopathology* **9**: 83-84 [Summary in Rev. Pl. Path. **58** (1979): 2440].

Harper, F.R. and Bergen, P. 1976. A powdery mildew on sugar beet in Alberta. *Canadian Plant Disease Survey* **56**: 48-52.

Jennrich, R. and Sampson, P. 1985. Stepwise discriminant analysis. pp. 519-537. In: BMDP Statistical Software Manual. - University of California Press, Berkeley.

Johnson, R.L. and Schall, R.A. 1989. Early detection of new pests. pp 105-116. In: *Plant Protection and Quarantine*. Vol III (ed. R.P. Kahn). CRC Press, Inc., Boca Raton, USA.

Jones, R.K., Barnes, L.W., Gonzalez, C.F., Leach, J.E., Alvarez, A.M., and Benedict, A.A. 1989. Identification of low-virulence strains of *Xanthomonas campestris* pv. *oryzae* from rice in the United States. *Phytopathology* **79**: 984-990.

Kaiser, W.J. and Muehlbauer, F.G. 1984. Occurence of *Ascochyta rabiei* on imported chickpeas in Eastern Washington. *Phytopathology* **74**: 1139.

Kranz, J., Mogk, M. and Stumpf, A. 1973. EPIVEN - ein Simulator für Apfelschorf. *Z. Pflanzenkr. Pflanzenschutz* **80**: 181-187.

Kranz, J. and Royle, D.J. 1978. Perspectives in mathematical modelling of plant disease epidemics. pp. 111-120. In: *Plant Disease Epidemiology* (eds. P.R. Scott and A. Bainbridge). Blackwell Scientific Publications, Oxford, UK.

Leppik, E.E. 1964. Mapping the world distribution of seed-borne pathogens. In: *Proceedings of the International Seed Testing Association* **29**: 473-477.

Lucas, M.T., Dias, M.R. de S. and Sequeira, M. 1979. Some fungi on sugar beet in Portugal. *Agronomia Lusitana* **39**, 361-376. [Summary in *Rev. Pl. Path.* **59**: (1980): 5428].
Miller, P.R. 1969. Effect of environment of plant diseases. *Phytoprotection* **50**: 81-94.
Müller, J. 1982. *Selected Climatic Data for a Global Set of Standard Stations for Vegetation Science.* Dr. W. Junk Publishers, The Hague, The Netherlands.
Nene, Y.L. 1982. A review of Ascochyta blight of chickpea. *Trop. Pest Management* **28**: 61-70.
Nene, Y.L. and Reddy, M.V. 1987. Chickpea disease and their control. pp. 233-270. In: *The Chickpea* (eds. M.C. Saxena and K.B. Singh). CAB International, Oxon.
Polley, R,W. and Clarkson, J.D.S. 1978. Forecasting cereal disease epidemics. pp. 141-150 In: *Plant Disease Epidemiology* (eds. P.R. Scott and A. Bainbridge). Blackwell Scientific Publications, Oxford, UK.
Ruppel, E.G., Hills, F.J. and Mumford, D.L. 1975. Epidemiological observations on the sugarbeet powdery mildew epiphytotic in Western U.S.A. in 1974. *Plant Disease Reporter* **59**: 283-286.
Schneider, C.L. and Hogaboam, G.J. 1977. New occurrence of powdery mildew and curly top on sugarbeet im Michigan in 1975. *Plant Dissease Reporter* **61**: 88-89.
Schoulties, C.L., Civerolo, E.L., Miller, J.W., Stall, R.E., Krass, C.J., Poe, S.R., DuCharme, E.P. 1987. Citrus canker in Florida. *Plant Disease* **71**: 388-395.
V"rbanov, V.M. 1978. (Morphological features of *Erysiphe communis* Grev. f. *betae* Jacz., the pathogen of powdery mildew of sugar beet). *Rasteniev"dni Nauki* **15**: 156-163 [Summary in *Rev. Pl. Path.* **58**: (1979) 5063].
Waggoner, P.E. 1968. Weather and the rise and fall of fungi. pp. 45-66. In: *Biometeorology* (ed. W.P. Lowry). Oregon State Univ. Press, Corvallis, OR, USA.
Weltzien, H.C. 1967. Geopathologie der Pflanzen. *Z. Pflanzenkr.* **74**: 176-189.
Weltzien, H.C. 1978. Geophytopathology. pp 339-360. In: *Plant Disease - An Advanced Treatise*. Vol. II (eds. J.G. Horsfall and E.B. Cowling). Academic Press, New York, San Francisco, London,
Weltzien, H.C. 1981. Geographical distribution of downy mildews. pp 32-43. In: *The Downy Mildews* (ed. D.M.N. Spencer). Academic Press, London.
Weltzien, H.C. 1988. Use of geophytopathological information. pp. 237-242. In: *Experimental Techniques in Plant Disease Epidemiology* (eds. J. Kranz. and J. Rotem). Springer Verlag, Berlin, Heidelberg, New York, London, Paris, Tokyo.
Whitney, E.D. and Lewellen, R.T. 1976. Identification and distribution of races C1 and C2 of *Cercospora beticola* from sugarbeet. *Phytopathology* **66**: 1158-1160.
Young, H.C., Jr., Prescott, J.M., Saari, E.E. 1978. Role of disease monitoring in preventing epidemics. *Ann. Rev. Phytopathol.* **16**: 263-285.

Country Reports
Round Table Discussions
and
Summary

Country Reports

Algeria

Mr. S. Ambar reported that Algerian imports include potatoes, faba bean, peas and oats. Quarantine measures are applied according to Plant Protection Legislation No. 87-17, and are carried out by the National Plant Protection Institute (INPV) on all shipments arriving at ports of entry. Phytosanitary certificates and treatment of certain imports are required. In a recent limited survey, the presence of seed-borne diseases on wheat, barley, chickpea, peas, lentil and faba bean was confirmed.

Cyprus

Ms. M. Dimova-Aziz reported that seed production, certification and distribution of certified seed are undertaken by the seed production centre of the Department of Agriculture. Seed certification is based on field inspection and laboratory tests. There is a well-organized and equipped seed testing laboratory available. In general, minimal quantities of seeds of various vegetables are exported by individuals, while about 26 kg of true seeds of various crops (mainly vegetables) are imported each year. Ms. Dimova-Aziz stated that the inspection of exported agricultural products is carried out by the inspection service of the Ministry of Commerce and Industry, while imports are controlled by the Plant Protection Section of the Department of Agriculture. Importation of seed is only allowed by import permit. Specific quarantine regulations for the importation of lettuce, tomato, pea and bean seeds are in force. She added that there is no seed health testing laboratory in the country. In the Plant Protection Section, there are some facilities for testing fungi, bacteria and viruses, as well as for nematode and insect examination. As there is no quarantine laboratory, imports rely mainly on the phytosanitary certificate. Additional staff and training, as well as laboratory equipment, are required to conduct effective seed health testing.

Jordan

Mr. S.H. Mufadi indicated that Jordan has legislation and regulations governing plant quarantine. This legislation was established in 1974 and is applied to plants and plant products including seeds. Jordan imports seeds of all vegetables and field crops which are grown in the country. Import permits and phytosanitary certificates are required. There is no seed export from Jordan. Seeds, as well as other plants, are visually inspected upon arrival at the points of entry, but the Quarantine Department does not have the facilities or trained personnel to make thorough seed examination. Recently a plant quarantine laboratory was established, but it lacks supplies, materials and trained personnel. Jordan would welcome any assistance to strengthen capabilities on quarantine for seed. Chemically-treated seeds are accepted without testing.

Libya

Dr. M.M. Baraka informed the workshop that quarantine for seed in Libya is regulated by the Plant Protection Law of 1968 and Executive Legislation of 1971 (now under review). Import permits are required to import seed. Authorized ports of entry, kinds of imported and exported seeds, and quarantine pests are specified in the regulations. Phytosanitary certificates are required. Seeds are inspected upon arrival at ports of entry by dry seed examination only, due to lack of facilities and trained personnel. However, in Tripoli, whenever required, limited seed health testing for fungi, bacteria and nematodes is performed by a seed pathologist at the University. No facilities for viral and advanced bacterial seed health testing are present. Dr. Baraka added that there is a need for a well-equipped central seed health testing laboratory, general training for inspectors and technicians, advanced training for viral and bacterial seed health testing, and post-entry growing facilities.

Morocco

Mr. A. Boughdad described the Moroccan quarantine system and indicated that the responsibility for implementation of quarantine falls within the authority of the Department of Plant Protection of the Ministry of Agriculture. Phytosanitary controls are carried out at the ports of entry into Morocco (seaports, airports and border stations) as well as inside the country. In an attempt to ensure appropriate quarantine measures for seed, it is still necessary to improve and reinforce the present infrastructure with adequate equipment, training and documentation, as well as updating of regulations and undertaking pest risk analysis studies.

Oman

Mr. A.Y. Al Baloushi stated that the application of plant quarantine measures started in Oman with the issuance of Law No. 49 of 1977. This was followed by several ministerial decrees regulating the entry of plants and plant products, including a list of prohibited plants. He further informed the workshop that there is a modest laboratory to test import and export consignments. Testing is performed visually with the aid of hand lenses. A fumigation station exists at the port of Kabour and another at Risuil airport. Samples of all shipments are visually inspected and phytosanitary certificates issued. Mr. Al Baloushi added that seed importation is permitted, subject to previous issuance of a permit from the Ministry of Agriculture and Fisheries. Although seeds are accompanied by a phytosanitary certificate, they have to be visually inspected before import is allowed. He informed the workshop that seeds are imported from the USA, Holland, Japan, Taiwan and India. He concluded that a seed health laboratory and training of staff on seed health testing are still required.

Pakistan

Dr. S.A.J. Khan reported that imported seeds are inspected upon arrival by visual examination of dry seed and, if necessary, by the blotter method. Similar procedures are followed for exports. In both cases, the consignment is thoroughly examined before issuance of a phytosanitary certificate by the Plant Protection Department in Karachi. Pathogens declared under import restrictions or prohibitions list are: *Synchytrium endobioticum*, *Heterodera rostochiensis*, *Spongospora subterranea*, *Leptinotarsa decemlineata*, *Ceratocystis paradoxa*, *Fusarium oxysporum*, bunchy top virus, *Aphelenchoides cocophilus* and *Puccinia arachidis*. There is a seed health testing laboratory for domestic seed, imported seed and seed for export. Testing procedures are: blotter and agar plant methods (for fungi), nutrient agar plate method (for bacteria) and Baermann funnel technique (for nematodes). Inspection for viral diseases is based on symptoms, but there are no facilities for the examination and identification of viral and bacterial diseases. Two seed health testing laboratories are operational: one at CDRI in Karachi and the other at the Federal Seed Certification Department in Islamabad. Both are in need of further strengthening with research personnel and equipment, which is imperative for effective plant quarantine in Pakistan.

Sudan

Dr. M.M. Obeid explained that Sudan has five points of entry. Of these five air, sea and road points of entry into the country, plant quarantine operates only at Port Sudan on the Red Sea coast and at Khartoum airport. Trained staff for seed health testing is inadequate and some testing facilities are needed, particularly in the field of virology. At present there are no post-entry quarantine or fumigation facilities. Seeds imported commercially are wheat, haricot beans, rice, lentil, alfalfa, sunflower, vegetables and ornamentals. Seeds exported include groundnut, sesame, sorghum, cotton, caster beans, vegetables and fruit. An import permit is required for seed importation. The import of cotton seeds is prohibited. Seeds are inspected upon arrival. There are no seed health testing laboratories for imported and exported seeds. Standard blotter test and dilution plating methods for bacteria are the only methods used. No antisera are available for virus testing.

Syria

Mr. A. Mahmoud reported that in 1991 new plant quarantine regulations were issued to supersede those of 1961. Annexed to these regulations were lists of quarantine significant pests prohibited for entry into the country or allowed with a certain tolerance. He pointed out that Syria has 19 plant quarantine entry points, each of which has basic facilities (binoculars, lenses, etc.) for inspection of consignments of plants and plant products, and is staffed by three to six plant quarantine officers. Random samples are withdrawn from seed consignments arriving at entry points, and these are forwarded to the seed health testing laboratory in Aleppo. The laboratory was established through assistance from ICARDA. It is adequately equipped and staffed for testing fungal and bacterial seed-borne pests. Due to lack of facilities required for testing pests of a viral nature, Syria relies entirely on phytosanitary certificates.

Turkey

Dr. G. Tuncer pointed out that in Turkey plant quarantine regulations are prepared and published by the Ministry of Agriculture and Rural Affairs, and implemented by the General Directorate of Plant Protection and Control. These regulations cover the export, import and transit of plants and plant products. Plants, including seeds, can be imported on condition that they are free from harmful organisms and conform to plant quarantine regulations. Seeds for planting have to be accompanied by a phytosanitary certificate from the Plant Protection Service of the country of origin in English, French or German. If the certificate is written in any other language, a Turkish translation has to be attached to the original certificate.

Yemen

Mr. A.A. Abdul Moghni stated that since at present there is a negligible level of seed production, more than 90 per cent of the required vegetable seeds are imported. There is no control on imported seeds. This results in illegal importation, losses due to more adaptable varieties, introduction of new pests and sale of poor quality seed. Mr. Abdul Moghni gave a few examples of introductions, including *Anguina tritici*, *Ditylenchus* sp., *Corynebacterium michiganense*, *Sclerotinia cepivorum* and *Verticillium* sp.), and said that seed health testing facilities and training on testing techniques are required in the country.

Conclusions from country reports

The following conclusions were drawn from the oral presentations of the participants from Algeria, Egypt, Cyprus, Jordan, Libya, Morocco, Oman, Pakistan, Sudan, Syria, Turkey and Yemen, as well on their replies to a questionnaire on quarantine for seed in their respective countries.

1. It is evident that all countries enact plant protection legislation and regulations, setting provisions for plant quarantine measures on both exports and imports of plants and plant products. Seeds are addressed by the provisions also covering other planting materials.

2. All participating countries are importers of seed (cereals, vegetables, legumes, fruit trees, fodder, ornamentals, etc.). An import permit is required for seed importations. Germplasm is provided mainly by International Agricultural Research Centres and some scientific institutions in developed countries. Most participants indicated that such germplasm is used for scientific purposes by research institutions, bypassing quarantine services.

3. Some countries of the region export seeds of a limited number of crops. No field inspection is exercised on such exports. Seed consignments for export are inspected at the port of exit and phytosanitary certificates issued.

4. Quarantine regulations include lists of prohibited and regulated entry quarantine pests. These lists concentrate on fungal pests, followed by bacteria, insects, nematodes, weeds and viral pests. Pests appearing in such lists include a large number of quality pests. In most cases, these lists are not regularly updated and in some countries they date back more than 20 years.

5. Upon arrival at the port of entry, seed consignments are subjected to visual inspection of the dry seed. If no significant pests are found, the consignments are released. Seeds showing symptoms or signs of infection are sent to laboratories of services or research institutions for seed health testing.

6. Facilities available at such institutions are mainly limited to testing for fungal pests. Testing for bacteria, viruses and nematodes is carried out in very few countries. None of the participating countries reported the existence of adequate facilities in quarantine services for seed health testing. Such situations result in almost complete reliance on phytosanitary certificates.

7. Most countries reported introductions and interceptions of quarantine pests through seed importations into their territory.

8. Personnel trained in seed health testing are not available in most countries, at least not in sufficient numbers. In most cases, they are not directly assigned to seed health tasting for quarantine purposes.

9. None of the plant quarantine stations is equipped with proper seed health testing facilities (equipment, glassware, chemicals, references, etc.) and, in most cases, they do not have staff trained in seed health testing.

10. Cooperation between the various institutions in participating countries on seed health testing for quarantine purposes appears to be unsatisfactory. Regional and sub-regional coordination on quarantine for seed is non-existent.

11. All participants indicated that, unless plant quarantine regulations are regularly revised, personnel trained in seed health testing, facilities provided, and national, sub-regional and regional coordination achieved, quarantine for seed will continue to be unsatisfactory and the threat of introduction and spread of devastating seed-borne pests will continue.

Round Table Discussions

The following points were raised in the round-table discussion:

Legislation and regulations

Seeds (i.e. true seeds for sowing) should be covered by the same phytosanitary regulations as other plants. These regulations should also apply to seeds for breeding and conservation (i.e. so-called "germplasm"), and should be considered in many instances to present a higher phytosanitary risk than commercial seeds. Grain (i.e. seeds for human consumption) should be considered as a separate category under the regulations and should not normally be subject to the same measures as seeds for sowing.

Lists of quarantine pests

The workshop gave its provisional agreement to a list of seed-borne pests (see the list) which may qualify for the status of quarantine pests in the Near East region. The pests in this list should now be subjected individually to pest risk analysis, following the recommendation of the FAO Technical Consultation among Regional Plant Protection Organizations. As a first phase, they should be adequately documented by data sheets in the FAO recommended format. As a second phase, an appropriate regional/international Expert Panel should review the data sheets and recommend quarantine pest status in appropriate cases. In due course, this project should be extended to all plants in order to draw up a complete quarantine list for the region.

It was noted that national regulations should specify only quarantine pests in the A list, for which the IPPC phytosanitary certificate specifies freedom, while quality pests should be listed, if at all, in a B list for which the certificate provides practical freedom.

The workshop reviewed the major crops imported as seeds into countries of the Near East region, taking into consideration which of the known pests might qualify as quarantine significant.

Training of staff in seed health testing

A strong need for training on seed health testing was perceived, and it was suggested that a regional centre for training phytosanitary inspectors in seed testing methods should be established. Such training should preferably concentrate on those crops which are attacked by quarantine pests for the region, rather than covering the full range of seed healthg testing.

List of potential seed-borne quarantine pests for the Near East region[1]

1. CEREALS

General insects pests of stored cereals

Prostephanus truncatus	*
Araecerus fasciculatus (mainly tropical)	?
Trogoderma granarium (probably in all Near East countries)	?

1.1 Small-grain temperate cereals (wheat/barley/oats)

Tilletia indica	*
Tilletia controversa	*
Urocystis agropyri	*
Tilletia caries and *T. foetida* (probably in all Near East countries)	?
Phaeosphaeria nodorum (*Septoria nodorum*) (favoured by cool Atlantic conditions only)	?
Gibberella zeae (*Fusarium graminearum*) (probably in all Near East countries) (N.B. also on maize)	?
Alternaria triticina (too little information on importance or distribution)	?
Barley strip mosaic hordeivirus (present in some Near East countries; status in others unknown)	?
Heterodera avenae (not a very important pest; distribution not known)	?

1.2 Rice

Xanthomonas campestris pv. *oryzae*	*
Xanthomonas campestris pv. *oryzicola*	*
Aphelenchoides besseyi	*

[1] * = quarantine status practically certain
? = quarantine status doubtful for reason given

.....cont.

Pyricularia oryzae ?
(probably qualifies, but well-distributed in some Mediterranean countries)

Sphaerulina oryzina (*Cercospora janseana*) ?
(tropical pest; status and potential in Near East uncertain)

Tilletia barclayana ?
(potential importance for, and status in, Near East uncertain)

1.3 Maize

Erwinia stewartii *

Stenocarpella maydis & *S. macrospora* *
(= *Diplodia*)

Setosphaeria turcica ?
(present in some Mediterranean countries)

Cochliobolus carbonum ?
(present in some Mediterranean countries)
(N.B. on EPPO list, but debatable)

Pernosclerospora philippinensis *

Pernosclerospora sorghi &
Sclerophthora macrospora ?
(present in some Near East countries, status needs reviw)

Gibberella zeae (*Fusarium graminearum*) ?
(probably in all Near East countries)
(N.B. also in wheat)

2. LEGUMES

2.1 Faba bean

Ditylenchus dipsaci *

Heterodera goettingiana ?
(is it important anywhere or seed-borne?)

2.2 Chickpea

Fusarium oxysporum f.sp. *ciceri* *

Heterodera ciceri ?
(is it important anywhere or seed-borne?)

.....cont.

2.3 Phaseolus beans

Phaeoisariopsis griseola *

Xanthomonas campestris pv. *phaseoli* *

Curtobacterium flaccumfaciens pv. *flaccumfaciens* ?
(a traditional quarantine pest which is on EPPO list, but is it important anywhere?)

2.4 Peas

Pseudomonas syringae pv. *pisi* ?
(potential importance for, and status in, Near East uncertain)

2.5 Groundnut

Peanut stripe potyvirus ?
Peanut clump furovirus ?
(both definitely not in Near East, but potential importance needs to be evaluated)

Ditylenchus destructor ?
(possibly widespread in Near East; concerns many other hosts)

2.6 Soybean

(little grown in Near East as yet, so most pests absent; a whole group needs to be evaluated for potential in Near East)

Phytophthora megasperma f.sp. *glycinea* *

Phialophora gregata *

Diaporthe phaseolorum complex *
(*Phomopsis sojae*)

Heterodera glycines *
(only marginally seed-borne, as a contaminant, but a serious pest)

Pseudomonas syringae pv. *glycines* *

Peronospora manshurica ?
(widespread in Europe, possibly in Near East)

Seed-borne viruses of soybean (to be reviewed as a group for importance and potential in Near East) ?

Septoria glycines ?
(mainly tropical)

.....cont.

Cercospora kikuchiana ?
(mainly tropical)

2.7 Pests of legumes in general

Seed-borne insects (e.g. bruchids, *Zabrotes*) ?
(some species may deserve quarantine status)

Striga spp. ?
(mainly tropical; potential in Near East uncertain)

3. COTTON

Anthonomus grandis *
(marginally seed-borne, contaminant only, but very serious pest)

Glomerella gossypii *

4. SUNFLOWER

Plasmopara halstedii *

Diaporthe helianthi ?
(importance for Near East needs evaluation)

5. SESAME

Xanthomonas campestris pv. *sesami*
and *Pseudomonas syringae* pv. *sesami* ?
(both are present in parts of the Near East, status needs review)

6. VEGETABLES

6.1 Tomatoes and Capsicums

Clavibacter michiganensis subsp. *michiganensis* *

Xanthomonas campestris pv. *vesicatoria* ?
(possibly well distributed in Near East)

Colletotrichum coccodes ?
(importance for, and status in, Near East uncertain)

Tomato bushy stunt virus ?
(importance for, and status in Near East uncertain)

6.2 Cucurbits

Squash mosaic comovirus *

Pseudomonas syringae pv. *lachyrmans* ?
(importance for, and status in, Near East uncertain)

.....cont.

Colletotrichum lagenarium ?
(importance for, and status in, Near East uncertain)

6.3 Onions

Urocystis cepulae ?
(importance for, and status in, Near East uncertain)

Colletotrichum circinonce ?
(importance for, and status in, Near East uncertain)

Laboratory facilities for seed health testing

It was recognized that most existing laboratories in the region are inadequately equipped. Detailed advice on the equipment needed for a simple seed testing laboratory was requested. The organizers of the workshop were invited to review the methods presented to the workshop, and to propose appropriate methods and detection equipment (preferably simple and inexpensive) on the basis of the list of potential quarantine pests prepared during the meeting. It was recognized that some necessary detection and identification methods cannot be performed by a simple laboratory, so individual cases may have to be referred to more specialized laboratories.

Library facilities (books, manuals, journals, etc.)

It was noted that the relevant books and journals on seed testing could be divided into those that should be held in a central library to which seed testing staff have access, and a relatively limited number of books and manuals which are essential for daily laboratory work. Since the staff concerned speak languages other than English (French and German), relevant books and manuals in such languages should also be specified.

Up-to-date information exchange between phytosanitary inspectors in different countries requires a central clearing house, and could be provided by the Secretariat of an RPPO.

Recommendations

In the light of the presentations made and the discussions held during the workshop, the participants from the various countries of the region expressed concern about the inefficiency of quarantine for seed in their countries in particular and in the Near East region in general. They stressed the need to strengthen quarantine for seed by ensuring the availability of adequate facilities for seed health testing; training of quarantine inspectors and technical staff on seed inspection, sampling and health testing; smooth flow and exchange of information; regular revision of regulations; and improvement of regional coordination and cooperation.

The workshop recommended that:

1. Countries of the Near East region should take the necessary steps to upgrade their quarantine services in terms of regulations, facilities, training and information available to enable quarantine authorities to exercise efficient quarantine measures on imports and exports of seed.

2. The workshop organizers prepare a project proposal on strengthening national and regional capabilities on quarantine for seed and submit such a proposal to potential donors for funding.

The recommendations made during the workshop were unanimously endorsed by all participants. Drs. Taher and Saxena expressed their thanks to all those who had contributed to the workshop, whether from international or regional organizations, scientific institutions or participating countries. They concluded by indicating that it is essential that the recommendations of the workshop, whether addressed to Near East countries or the organizers, should be followed up closely and implemented.

Summary

International and Regional Cooperation in Plant Quarantine

In his presentation on the **International Plant Protection Convention**, Dr. M.M. Taher of FAO pointed out that the spread of pests and diseases via the movement of plants in international trade has historically resulted in crop losses leading to shortage of production, famine, instability of agro-industries and the application of expensive control measures. Such a situation caused international concern, which led to the establishment of the 1881 International Convention against *Phylloxera vastatrix* and the 1929 International Convention of the protection of plants. These conventions were superseded in 1951 by the International Plant Protection Convention. This convention provides the framework for all legislation and regulations for export and import of plants and plant products, and hence provides all elements for fair trade. The purpose of the Convention is to ensure effective international cooperation to contain the introduction and spread of plant pests and diseases, particularly those of importance to trade. It includes provisions for supplementary agreements; establishment of national organizations for plant protection; issuance of phytosanitary certificates; requirements in relation to imports; international cooperation; establishment of regional plant protection organizations; and settlement of disputes. Dr. Taher indicated that the International Plant Protection Convention was revised in 1979 and the revised text was enforced in April 1991. The Food and Agriculture Organization of the United Nations is the depository and administrator of the Convention, which is currently adhered to by 96 countries. The Convention has contributed to strengthening national plant protection capabilities; enhancing regional and international cooperation; the establishment of information systems on pests and their distribution; minimizing the introduction and spread of plant pests; and alleviation of unnecessary phytosanitary measures in international trade.

Dr. E. Feliu of FAO presented the topic "**Quarantine for seed: Status, requirements and implications**". He explained that the international exchange of seeds and propagation material is necessary in order to sustain valuable research and breeding programmes aimed at improving and increasing crop production. However, he clarified that this poses a serous pest introduction and dissemination risk, unless appropriate safeguards are implemented. He added that, unfortunately, most quarantine services in developing countries lack some of the essential elements that would enable them to apply these safeguards and, therefore, national quarantine services need to be updated to a fully operational level (appropriate regulatory framework, facilities and staffing). They should also have access to up-to-date and reliable information on pests and regulations in other countries. While this is undertaken, all those concerned with the international movement of seeds must cooperate to ensure their safe movement. This cooperation must be present at national, regional and global levels as required by the International Plant Protection

Convention. Only in this way may the safe and expeditious movement of seeds be ascertained and screened.

EPPO's experience of quarantine for seed was presented by Dr. I.M. Smith of OEPP/EPPO, who explained that EPPO's basic phytosanitary strategy is to list quarantine pests and make specific quarantine requirements for each. Only a limited number of the pests in the EPPO A1 and A2 lists are seed-borne and these concern a fairly limited number of host plants (mainly cotton, maize, Phaseolus beans, rice, soybean, tomato and wheat), which in some cases concern relatively few member countries. Therefore, EPPO member countries require a Phytosanitary Certificate for seeds in only a few cases, but then make specific requirements. These are mostly for area freedom, field inspection of the seed crop, and seed health testing or treatment. Prohibition of seeds or post-entry quarantine procedures are seldom recommended. Ideally, inspection, treatment and test procedures should be specified. EPPO believes that better phytosanitary security is achieved by following an agreed testing procedure, with an implied tolerance, than by delivering certificates based on unspecified procedures.

Seed-borne Diseases of Economic Importance

Dr. O. Mamluk of ICARDA reported on **Seed-borne diseases of wheat and barley** in Near East countries and listed 19 seed-borne diseases (14 fungal, three bacterial, one viral and one nematode) on these crops. He indicated that seed-borne diseases of wheat and barley have been introduced to newly developed, irrigated areas in the region. Dr. Mamluk considered smuts the most important seed-borne diseases. Studies in genetic variation to smut pathogens or plant resistance are still limited in the region. He concluded that chemical seed treatment is very effective in controlling smuts, but the use of resistant cultivars still remains the most feasible control measure.

In relation to **Diseases of faba bean and lentil**, Dr. B. Bayaa of University of Aleppo, Syria explained that faba bean and lentil are affected by a number of seed-borne pathogens which are serious or potentially serious under favourable epidemiological conditions. The identification of such pathogens can be based on visual symptoms and/or cultural characteristics in nutrient media. He highlighted the epidemiological aspects of such diseases and provided the major methods for their control.

Dr. M.P. Haware of ICRISAT presented a paper on **Chickpea and groundnut seed-borne diseases of economic importance: Transmission, detection and control**. He explained that *Fusarium oxysporum* f.sp. *ciceri* and *Ascochyta rabiei* are economically important seed-borne pathogens of chickpea. *Sclerotinia minor*, *Pseudomonas solanacearum*, *Ditylenchus destructor*, peanut mottle virus, peanut stripe virus, peanut clump virus, peanut stunt virus and cucumber mosaic virus are seed-transmitted pathogens of groundnut. He discussed transmission and detection of these pathogens, and indicated that control methods have been developed for them but that additional research on some of the seed-transmitted pathogens of chickpea and groundnut is needed.

Dr. J. Taylor of Horticulture Research International, Wellesbourne, U.K., reporting on **Seed-borne bacterial diseases of significance to vegetable crops**, indicated that more than 40 crops of economic importance are grown in temperate regions. They are affected by a range of important seed-borne diseases caused by fungi, bacteria and viruses. Examples of these infections were discussed for brassicae (*Phoma lingam*, *Alternaria brassicae*, *Xanthomonas campestris* pv. *campestris*); bean (*Colletotrichum lindemuthianum*, *Pseudomonas syringae* pv. *phaseolicola*, bean common mosaic virus); lettuce (lettuce mosaic virus); celery (*Septoria apiicola*); tomato (*Clavibacter michiganensis* subsp. *michiganensis*); onion (*Botrytis allii*); pea (*Ascochyta pisi*, *Pseudomonas syringae* pv. *pisi*); cucurbits (zucchini yellow mosaic virus) and many different vegetables (alfalfa mosaic virus).

Seed Exchange

Dr. M. Diekmann of ICARDA described the **Procedures followed in exchange of cereals, food legumes and fodder seeds at ICARDA** and the problems involved. Germplasm for distribution to cooperators abroad is produced in locations with low disease pressure. During the growth period fields are regularly inspected and, if necessary, pesticides are applied. Before dispatch, laboratory tests are carried out in accordance with the importing country's regulations and the seeds are treated with broad-spectrum fungicides. Newly introduced germplasm, after having passed the Syrian quarantine, is subject to post-entry quarantine by planting in an isolation area and regular field inspections. Problems encountered are: quantities available for testing may be insufficient; lack of non-destructive test methods; testing of treated seeds may be difficult; field inspection is time-consuming because thousands of lines need to be inspected individually. A list of prohibited pests from cooperating countries and more information on the pathology of wild relatives are required.

Dr. M.P. Haware reported on **Germplasm exchange and plant quarantine systems at ICRISAT**, on behalf of Drs. Mengesha and Joshi. He pointed out that ICRISAT has assembled and conserved over 100,000 accessions of sorghum, pearl millet, chickpea, pigeonpea, groundnut and minor millets. Over half-a-million samples of germplasm have been distributed for crop improvement programmes in 128 countries over the past 15 years. To safeguard against the possible entry of exotic pests, disease and weeds during exchange, ICRISAT strictly adheres to international guidelines and specific national quarantine regulations. To export seed from ICRISAT, healthy seed produced during post-rainy season is conserved in the gene bank. Healthy seed is sent to the recipient together with a phytosanitary certificate. All seeds imported for ICRISAT are first received at the port of entry. They are forwarded to the National Bureau of Plant Genetic Resources for extensive examination and specific treatment as prescribed by the International Seed Testing Association and FAO's International Plant Protection Convention of 1951. Later, seeds are planted in the Post Entry Quarantine Isolation Area at ICRISAT Center for field examination against contamination with pests. Healthy and clean seeds are harvested and delivered to ICRISAT. Though 165,925 samples were introduced to ICRISAT from 95 countries and 804,420 were dispatched to 147 countries from ICRISAT, no pest pathogen or weed was introduced or exported from ICRISAT to any country. This demonstrates that the present system followed by ICRISAT is safe and reliable.

Dr. A. Abdelmonem of Agricultural Research Center, Giza, Egypt briefly described **Seed Movement and Quarantine Measures in Egypt** and provided examples of successful interception and prevention of introduction of seed-borne pathogens in seed movement from Near East countries. Provisional regulations, phytosanitary certificates, categories of seed-borne pests as quarantine objectives were discussed. Dr. Abdelonem informed the workshop that seed health testing procedures are practised to detect seed-borne fungi, bacteria, viruses and nematodes. He further explained that virus detection methods for testing treated seed and shortage of quarantine service staff trained in seed pathology are primary constraints, affecting proper quarantine control of seed movement. He also stressed the significance of rapid and reliable detection methods and more up-to-date information on the geographical distribution of seed-borne pathogens in the Near East in ensuring appropriate quarantine controls for the movement of seed.

On behalf of Dr. R. Ikin of FAO, Dr. E. Feliu presented a paper entitled **Precautions directed against the introduction of pathogens and pests into uninfected areas**. He explained that effective plant quarantine procedures are necessary to eliminate or minimize the risks associated with the international exchange of seeds. He said that an effective mechanism needs to be in place in order to adequately identify those pests that are not present in the importing country and which are a substantial economic risk. The classification of seed imports into categories of low to high risk should result from this assessment. Dr. Feliu added that, at a time when phytosanitary regulations are required to meet the "phytosanitary need", testing is becoming increasingly important. He further informed the workshop that the high risk seed categories require growing and observation under post-entry quarantine conditions. Particular care must be given to these categories of seed. It has been demonstrated that usually simple, durable structures are sufficient to provide quarantine security, if well-staffed and managed. This is particularly true with respect to developing countries where the construction of highly sophisticated structures has proved impractical, given the lack of sufficient resources and technical capabilities to operate and maintain them. The staff of these facilities should include a skilled horticulturist to ensure and maintain the appropriate cultivation of the plants. Other specialists need not be constantly present at the facility, but may conduct regular visits to observe the plants whenever necessary. Ideally, these facilities should be located near established research/diagnostic institutions for technical support when needed.

Pests of Quarantine Significance

Mr. H.J. Hansen of Danish Government Institute of Seed Pathology for Developing Countries, Denmark discussed **Seed-borne fungi of quarantine significance**. These pathogens, when introduced into new areas, are able to cause considerable damage to the crop. Regulations regarding import of seed material which can carry such fungi should be based on a pest risk analysis. Import should be allowed on condition of freedom from quarantine significant fungi (zero tolerance required). On a national basis, these fungi are listed in category A. On a regional basis, they fall into the quarantine category A1 (fungus not present in the region) or A2 (pathogen present in some countries of the region).

In his presentation on **Bacterial pests of quarantine significance**, Dr. J. Taylor explained the problems associated with the quarantine and quality considerations in relation to two seed-borne bacterial diseases, *Pseudomonas syringae* pv. *phaseolicola* and *P. s.* pv. *pisi*. The latter is restricted to Europe while the former, which is of worldwide distribution, is only of quality significance. The similarities between the pathogens were noted; both have a complex structure and their distribution is essentially influenced by the presence of resistant genes in their hosts.

The importance of **Viruses as quarantine pests** was stressed by Dr. L. Bos of Research Institute for Plant Protection, Wageningen, Netherlands, who indicated that viruses play a very special role with respect to seed health. He mentioned that viruses which are transmitted via the embryo stay infective in seed as long as these remain viable. They are usually symptomless in seeds, often hard to detect and, at the same time, they usually cannot be removed from seed by disinfection. Dr. Bos explained the mechanisms by which viruses become seed-borne and seed-transmitted. Whether a seed-borne virus should be assigned quarantine status is often hard to decide. Seed testing methods for virus detection are usually destructive. Virus-free seed can be obtained from plants that proved virus-free in seed samples of germplasm sown under quarantine. Large quantities of commercial seed which cannot be handled by quarantine must be propagated in the open with expected risk of re-infection and checked for health by simple testing methods. Requirements with respect to virus detection in seed (from germplasm via breeders' lines to commercial seed) must be realistic and are often a matter of compromise.

Dr. G. Caubel of INRA Zoologie, France presented a paper on **Seed-borne nematode species of quarantine significance**. He indicated that several nematode genera are known to be seed-borne: most important in agriculture are *Anguina*, *Aphelenchoides* and *Ditylenchus*. Many crop species may be infested, including rice, wheat, legumes, groundnut and onion. In some cases, seed infection is the dominant means of survival. Investigations have shown the importance of infested seed stocks, but more attention should be paid to *Vicia* bean infested with the stem nematode, *Ditylenchus dipsaci*. Dr. Caubel also indicated that nematodes could adhere to the seed surface, with plant debris, within the cotyledons or as solid masses. They survive adverse environments; some well- adapted larval stages remain alive in the anhydrobiotic state and, with an increase in moisture, they become active and invade the seedlings. Inoculum is either dispersed in a stock with debris or very concentrated in a seed. In field conditions, establishment of disease appears primarily in isolated plants or in circular patches. The best control measures consist of prevention of spread by production of healthy seeds in a certification scheme, good detection methods, seed cleaning or the use of physical agents.

In his presentation on **Insect pests of quarantine significance to seed**, Dr. K. Richter of University of Leipzig, germany indicated that seeds are subject to attack by many species of insect. According to their damage potential, grain storage insects can be classified into two groups, primary and secondary pests. Primary pests that develop from an egg (or larva) to the adult stage inside the seeds are of great importance for seed quality. They are able to build up high density populations by increasing only the level of internal infestation without visible symptoms. Subjects of quarantine significance to seed may be listed as: *Prostephanus truncatus*, *Araecerus fasciculatus* and *Zabrotes subfasciatus*, according to their climatic requirement and host plant range. Population development can be inhibited by cold storage of the seed at 12-15° C. In relation to the use of post-harvest insecticides, attention must be paid to the development of pesticide resistance.

Seed Health Testing

In his presentation on **Samplings procedures**, Mr. H.J. Hansen emphasized that the sample drawn from a consignment should be a "representative" sample and should be further divided in the laboratory using different mechanical dividers for obtaining "working" samples. Mr. Hansen described the different devices which are used in obtaining well-mixed sub-samples. He explained that, in germplasm exchange, all seed must be tested for harmful pests and that they should be tested by non-destructive testing methods.

Seed health testing methods used in detecting fungi were also described by Mr. H.J. Hansen. He presented commonly used methods, such as inspection of dry seed, examination of surface washings (washing test), incubation methods (blotter method and agar plate method), embryo count method and seedling symptom test. Advantages and disadvantages of these methods were discussed, together with their application in screening seed samples in a plant quarantine laboratory.

In his presentation on **Seed health testing for bacteria**, Dr. J. Taylor described the general methods of testing which involve 1) extraction, 2) isolation and 3) identification. The quantification of seed-borne infection was also discussed. A rapid identification test of particular use in seed testing, based on the use of specific antisera conjugated to *Staphylococcus aureus* was presented. Dr. Taylor believed that this technique would have major application in quarantine laboratories.

Dr. D. Spire of INRA, France introduced **Novel approaches in testing for seed-borne fungi**. He gave examples of cases where it was possible to obtain a more specific, sensitive and early diagnosis of certain fungi. Some new experiments have been carried out with fungus protein for the purpose of producing specific antisera. He said that it is necessary to choose those proteins which are specific to the fungus studied. If there is only one fungus in a host plant, it is possible to obtain total protein contents in an antiserum whose antibodies will permit diagnosis of the pathogen in the plant or seed. Examples were given of fungal pathogens where antisera have been prepared. Dr. Spire pointed out that the use of antisera in diagnosing bacteria and viruses is common and he hoped that several antisera for fungi would soon be available.

Dr. K. Makkouk of ICARDA indicated that **Testing procedures for seed-borne viruses** should be simple, economical, sensitive and rapid. He pointed out that growing-on and infectivity tests have been used for a long time and have proved useful for detecting seed-borne viruses. For such tests to be reliable, they require a glasshouse facility where temperature and light intensity can be maintained at optimum levels for symptom development. More recently, a number of serological tests have been developed with a higher level of sensitivity than earlier tests. These are double antibody sandwich ELISA carried out on either polystyrene plates or nitrocellulose membranes. The sensitivity of this test can also be greatly increased by using amplification mixtures which allow the detection of one picogram virus per millilitre of seed extract. Virobacterial agglutination test is also a new procedure which is simple and rapid. Dr. Makkouk indicated that the Virology Laboratory at ICARDA is prepared to provide interested laboratories in Near East countries with ELISA kits using the above technique for the detection of a number of cereal and legume seed-borne viruses.

Dr. G. Caubel gave an overview of **Seed health testing for seed-borne nematodes** and emphasized that extraction of nematodes from seeds should only be carried out in cases where symptoms may not be recognized on seeds. He gave details of common methods, e.g. Baermann funnel technique, mist extraction system, centrifugal flotation method, biological testing and X-ray radiography. Dr. Caubel emphasized the main points which help in distinguishing parasitic from saprozoic nematodes.

Discussing **Seed health testing for insect pests**, Dr. K. Richter pointed out that the ultimate success of any strategy for pest control and quarantine requirements depends on the effectiveness of the methods used for the detection of the pests. Dr. Richter provided an overview of the most common methods for monitoring, sampling and detection of free-living stages as well as for internal infestation. He indicated that it is important for modification of detection methods and their adaptation to specific seed-pest combinations. One example is the improvement of the flotation method with salt solutions resulting in higher detection capacity. Summarizing the general information and experience gained in the Near East region, a recommendable combination of methods for monitoring and detection of all relevant developmental stages of insects was introduced for discussion.

Mr. H.J. Hansen presented **Methods used in testing chemically-treated seed**. He explained the procedures whereby laboratory testing can determine whether a seed consignment has received the right dosage of a chemical and whether individual seeds have in general received a uniform coating of fungicides (especially contact fungicides). Mr. Hansen said that chemically-treated seed samples must be handled with care and caution, and special provisions must be made in testing laboratories where seed packets can be opened. Pregnant women must not be allowed to work with chemically-treated seed.

In his presentation on **Seed health testing laboratories: Facilities and management**, Dr. S.B. Mathur of Danish Government Institute of Seed Pathology for Developing Countries, Denmark told that seed health refers to the presence or absence of disease or damage-causing organisms such as fungi, bacteria, viruses, nematodes and insects. This definition, as formulated by the International Seed Testing Association, fits very well the work of plant quarantine stations. He informed that testing techniques are available to detect organisms belonging to these groups, but a well-equipped laboratory plus trained staff are necessary to perform effective and efficient testing.

Control and special topic

Dr. E. Feliu gave a presentation on **Chemical and physical seed treatments**. He stated that the aim of plant quarantine treatments is to eliminate plant pests without significantly affecting or injuring the host commodity. Unfortunately, except for pests like insects and nematodes, there are no plant quarantine treatments for seeds that would provide full quarantine security. Current quarantine treatments are divided into two major categories: physical and chemical treatments. Except for some fumigants such as methyl bromide, chemical treatments are only effective against a limited number of pests, may induce resistance and leave chemical residues on the commodity. Physical treatments, on the contrary, are effective against a broad spectrum of pests, leave no residues on the host and do not induce resistance in insects and other pests. They are also relatively easy to apply and are safer to man, animals and the environment. The choice of a particular treatment to be used must take into consideration the pest/host relationship, availability of facilities and the technical capabilities in the country.

In connection with the theme **Elimination of infection from valuable seed germplasm**, Dr. P.E. Kyriakopoulou of Benaki Phytopathological Institute, Athens, Greece described the various ways of freeing seed germplasm from fungal, bacterial, viral and nematode infection. Methods are mechanical or physical and include removal of infected seeds or concomitant contamination, washing or surface sterilization, heat treatment, gamma irradiation, ageing, and the use of protectant or systemic fungicides. For the inspection of seed-borne viruses, the use of sensitive detection techniques is the most effective procedure. Tissue culture is not as necessary as with vegetative propagative material, but it may also be used. Dr. Kyriakopoulou concluded that valuable seed germplasm must be completely pathogen-free.

Dr. M. Diekmann presented results of **Geophytopathology of some major seed-borne diseases**. Geophytopathology has two aspects: mapping of disease distribution and risk assessment. Based on the distribution of host plants and pathogens, and on climatological data, an assessment of the likelihood of the establishment of a pathogen in new areas can be made. Examples for analysis of Ascochyta blight of chickpea and bacterial leaf blight (BLB) of rice were given. By comparing climatic data from locations where these diseases occur with those from areas where the hosts are grown but the diseases not reported, the discriminant factors could be identified. Based on a linear discriminant function with temperature and precipitation variables, new locations could be classified in "risk" or "non-risk" groups.

Lists of participants, Instructors and Organizing Committee

List of participants

Algeria

Ambar, Said
I.T.G.C.
1 Rue Pasteur BP 16
El Harrach — Telex: 64130 IGRAC
Telephone: 764431

Cyprus

Dimova-Aziz, Maria
Department of Agriculture
Ministry of Agriculture and Natural Resources
Nicosia — Telex: 4660 MINAGRI CY
Telephone: 30-2273

Egypt

Abdelmonem, Abdalla
Agricultural Research Center
Plant Pathology Research Institute
Gamaa Street
P.O. Box 12619 Giza — Telefax: 02722609
Telephone: 723442-723906-2588159

Libya

Baraka, Mokhtar Mohamed
Secretariat for Agricultural Reclamation & Land Development
P.O. Box 351 Tripoli — Telex: 20105
Telefax: (021) 603449
Telephone: (021) 39141/48

Morocco

Boughdad, Ahmed
ENA (MARA)
BP S/40 Meknès — Telex: 42154 ENAMEK
Telefax: 212 (05) 51 16 55
Telephone: (05) 523886, 522389, 511887/86

Jordan

Mufadi, Said Hamed
Department of Plant Protection
Ministry of Agriculture
P.O. Box 2099, 961043, 961044
Amman — Telex: 24176 AGRI JO
Telefax: 686310
Telephone: 726201/202

Oman

Al Baloushi, Ahmed Youssef
P.O. Box 467
Muscat — Telephone: 696287 - 696300

Pakistan

Khan, Sardar Abdul Jamil
Pakistan Agricultural Research Council
C.D.R.I.
Karachi Telephone: 474063

Sudan

Obeid, Mahmoud Mustafa
Plant Protection Directorate
P.O. Box 14 Khartoum North, or P.O. Box 624 Omdurman
Telephone: 612666

Syria

Assi, Walid
Plant Protection Division
Agriculture Directorate
Aleppo Telephone: 228920

Attar, Loutfi
Plant Protection Division
Agriculture Directorate
Aleppo Telephone: 214821

El-Ahmed, Ahmed
ICARDA
P.O. Box 5466
Aleppo Telex: 331206, 331208 ICARDA SY
Telefax: 225105, 213490
Telephone: 213433, 235221

Hurh, Ahmed
Plant Protection Division
Agriculture Directorate
Aleppo Telephone: 440150

Mahmoud, Ali
Plant Quarantine Division
Agricultural and Agrarian Reform Ministry
Damascus Telephone: 721808

Youssef, Gabriel
Agricultural Science Research Center
P.O. Box 713 Douma
Damascus Telephone: 750401

Turkey

Tuncer, G.
Plant Protection Research Institute
P.O. Box 49 Yenimahalle
Ankara Telefax: 90 (4) 3151531
Telephone: 90 (4) 3445993

Yemen

Abdo, M. Abdul Aziz
Ministry of Agriculture, (Plant Protection)
Aden — Telefax: 009672 228064
Telephone: 33262, 32611

Abdul Moghni, Abbas Ali
Plant Protection Project
P.O. Box 26 Sana'a — Telefax: 009672-228064
Telephone: 009672 227972

List of Instructors

Abdelmonem, Abdalla
Agricultural Research Center
Plant Pathology Research Institute
Gamaa Street
P.O. Box 12619 Giza
Egypt
Telefax: 02722609
Telephone: 723442-723906-2588159-2820440

Bayaa, Bassam
Faculty of Agriculture
University of Aleppo
P.O. Box 6452/ P.O. Box 5466
Aleppo
Syria
Telephone: 231244

Bos, L.
Research Institute for Plant Protection (IPO-DLO)
P.O. Box 9060
6700 GW Wageningen
Netherlands
Telex: 45888 NTAS NL
Telefax 08370-10113
Telephone: 08370-76002,76278

Caubel, Georges
INRA Zoologie
35650 Le Rheu
France
Telex: 740060F INRA LR
Telefax: 99285250
Telephone: 99645287

Diekmann, Marlene
ICARDA
P.O. Box 5466
Aleppo
Syria
Telex: 331206, 331208 ICARDA SY
Telefax: 225105, 213490
Telephone: 213433, 235221

Feliu, Edwin
FAO
Via delle Terme di Caracalla
00100 Rome, Italy
Telex: 610181 FAO I
Telefax: 5782610/57973152
Telephone: (06) 57976803

Hansen, Henrik J.
Danish Government Institute of Seed Pathology for Developing Countries
Ryvangs Allè 78
2900 Hellerup
Denmark
Telex: 16949 SEPATH DK
Telefax: 45 31 620212
Telephone: 45 31 621213

Haware, M.P
ICRISAT
PO, Patancheru 502324
Andhra Pradesh
India
Telex: 0422-02300, 0425-6366
Telefax: (91-842) 241239
Telephone: 224016-20 (ext. 380)

Kyriakopoulou, P.E.
Benaki Phytopathological Institute
14561 Kiphissia
Athens, Greece
Telefax: 01-8077506
Telephone: 01-8077829

Makkouk, Khaled
ICARDA
P.O. Box 5466
Aleppo, Syria
Telex: 331206, 331208 ICARDA SY
Telefax: 225105, 213490
Telephone: 213433, 235221

Mamluk, Omar
ICARDA
P.O. Box 5466
Aleppo, Syria
Telex: 331206, 331208 ICARDA SY
Telefax: 225105, 213490
Telephone: 213433, 235221

Mathur, S.B.
Danish Government Institute of Seed Pathology for Developi
Ryvangs Allè 78
2900 Hellerup, Denmark
Telex: 16949 SEPATH DK
Telefax: 45 31 620212
Telephone: 45 31 621213

Richter, Klaus
University of Leipzig
Institute of Tropical Agriculture
Fichtestr. 28
D-7030 Leipzig, Germany
Telephone: 80326

Smith, I.M.
OEPP/EPPO
1, rue Le Nôtre
75016 Paris, France
Telex: 614148
Telefax: 1-42-24-89-43
Telephone: 1-45-20-77-94

Spire, Didier
INRA France
Pathologie Végétale
Route de Saint-Cyr
78000 Versailles, France
Telex: 695269F INRAVER
Telefax: (1) 30833459
Telephone: (1) 30833220

Taher, Mahmoud Mohamed
FAO Regional Office for the Near East
P.O. Box 2223
Dokki
Cairo, Egypt
Telex: 21055 FAONE UN
Telefax: 3495981
Telephone: 3497184/702229

Taylor, John
Crop Protection Department
Horticulture Research International
Wellesbourne
Warks. CV35 9EF, U.K.
Telex: 311827 MANCEN G
Telefax: (789) 470552

Organizing Committee

Diekmann, Marlene
ICARDA
P.O. Box 5466
Aleppo
Syria
Telex: 331206, 331208 ICARDA SY
Telefax: 225105, 213490
Telephone: 213433, 235221

Makkouk, Khaled
ICARDA
P.O. Box 5466
Aleppo
Syria
Telex: 331206, 331208 ICARDA SY
Telefax: 225105, 213490
Telephone: 213433, 235221

Mathur, S.B.
Danish Government Institute of Seed Pathology for Developing Countries
Ryvangs Allè 78
2900 Hellerup
Denmark
Telex: 16949 SEPATH DK
Telefax: 4531 620212
Telephone: 4531 621213

Taher, Mahmoud Mohamed
FAO Regional Office for the Near East
P.O. Box 2223
Dokki
Cairo
Egypt
Telex: 21055 FAONE UN
Telefax: 3495981
Telephone: 3497184/702229

FAO TECHNICAL PAPERS

FAO PLANT PRODUCTION AND PROTECTION PAPERS

1	Horticulture: a select bibliography, 1976 (E)
2	Cotton specialists and research institutions in selected countries, 1976 (E)
3	Food legumes: distribution, adaptability and biology of yield, 1977 (E F S)
4	Soybean production in the tropics, 1977 (C E F S)
4 Rev.	1. Soybean production in the tropics (first revision), 1982 (E)
5	Les systèmes pastoraux sahéliens, 1977 (F)
6	Pest resistance to pesticides and crop loss assessment – Vol. 1, 1977 (E F S)
6/2	Pest resistance to pesticides and crop loss assessment – Vol. 2, 1979 (E F S)
6/3	Pest resistance to pesticides and crop loss assessment – Vol. 3, 1981 (E F S)
7	Rodent pest biology and control – Bibliography 1970–74, 1977 (E)
8	Tropical pasture seed production, 1979 (E F** S**)
9	Food legume crops: improvement and production, 1977 (E)
10	Pesticide residues in food, 1977 – Report, 1978 (E F S)
10 Rev.	Pesticide residues in food 1977 – Report, 1978 (E)
10 Sup.	Pesticide residues in food 1977 – Evaluations, 1978 (E)
11	Pesticide residues in food 1965-78 – Index and summary, 1978 (E F S)
12	Crop calendars, 1978 (E/F/S)
13	The use of FAO specifications for plant protection products, 1979 (E F S)
14	Guidelines for integrated control of rice insect pests, 1979 (Ar C E F S)
15	Pesticide residues in food 1978 – Report, 1979 (E F S)
15 Sup.	Pesticide residues in food 1978 – Evaluations, 1979 (E)
16	Rodenticides: analyses, specifications, formulations, 1979 (E F S)
17	Agrometeorological crop monitoring and forecasting, 1979 (C E F S)
18	Guidelines for integrated control of maize pests, 1979 (C E)
19	Elements of integrated control of sorghum pests, 1979 (E F S)
20	Pesticide residues in food 1979 – Report, 1980 (E F S)
20 Sup.	Pesticide residues in food 1979 – Evaluations, 1980 (E)
21	Recommended methods for measurement of pest resistance to pesticides, 1980 (E F)
22	China: multiple cropping and related crop production technology, 1980 (E)
23	China: development of olive production, 1980 (E)
24/1	Improvement and production of maize, sorghum and millet – Vol. 1. General principles, 1980 (E F)
24/2	Improvement and production of maize, sorghum and millet – Vol. 2. Breeding, agronomy and seed production, 1980 (E F)
25	*Prosopis tamarugo:* fodder tree for arid zones, 1981 (E F S)
26	Pesticide residues in food 1980 – Report, 1981 (E F S)
26 Sup.	Pesticide residues in food 1980 – Evaluations, 1981 (E)
27	Small-scale cash crop farming in South Asia, 1981 (E)
28	Second expert consultation on environmental criteria for registration of pesticides, 1981 (E F S)
29	Sesame: status and improvement, 1981 (E)
30	Palm tissue culture, 1981 (C E)

31	An eco-climatic classification of intertropical Africa, 1981 (E)
32	Weeds in tropical crops: selected abstracts, 1981 (E)
32 Sup.	1. Weeds in tropical crops: review of abstracts, 1982 (E)
33	Plant collecting and herbarium development, 1981 (E)
34	Improvement of nutritional quality of food crops, 1981 (C E)
35	Date production and protection, 1982 (Ar E)
36	El cultivo y la utilización del tarwi – *Lupinus mutabilis* Sweet, 1982 (S)
37	Pesticide residues in food 1981 – Report, 1982 (E F S)
38	Winged bean production in the tropics, 1982 (E)
39	Seeds, 1982 (E/F/S)
40	Rodent control in agriculture, 1982 (Ar C E F S)
41	Rice development and rainfed rice production, 1982 (E)
42	Pesticide residues in food 1981 – Evaluations, 1982 (E)
43	Manual on mushroom cultivation, 1983 (E F)
44	Improving weed management, 1984 (E F S)
45	Pocket computers in agrometeorology, 1983 (E)
46	Pesticide residues in food 1982 – Report, 1983 (E F S)
47	The sago palm, 1983 (E F)
48	Guidelines for integrated control of cotton pests, 1983 (Ar E F S)
49	Pesticide residues in food 1982 – Evaluations, 1983 (E)
50	International plant quarantine treatment manual, 1983 (C E)
51	Handbook on jute, 1983 (E)
52	The palmyrah palm: potential and perspectives, 1983 (E)
53/1	Selected medicinal plants, 1983 (E)
54	Manual of fumigation for insect control, 1984 (C E F S)
55	Breeding for durable disease and pest resistance, 1984 (C E)
56	Pesticide residues in food 1983 – Report, 1984 (E F S)
57	Coconut, tree of life, 1984 (E S)
58	Economic guidelines for crop pest control, 1984 (E F S)
59	Micropropagation of selected rootcrops, palms, citrus and ornamental species, 1984 (E)
60	Minimum requirements for receiving and maintaining tissue culture propagating material, 1985 (E F S)
61	Pesticide residues in food 1983 – Evaluations, 1985 (E)
62	Pesticide residues in food 1984 – Report, 1985 (E F S)
63	Manual of pest control for food security reserve grain stocks, 1985 (C E)
64	Contribution à l'écologie des aphides africains, 1985 (F)
65	Amélioration de la culture irriguée du riz des petits fermiers, 1985 (F)
66	Sesame and safflower: status and potentials, 1985 (E)
67	Pesticide residues in food 1984 – Evaluations, 1985 (E)
68	Pesticide residues in food 1985 – Report, 1986 (E F S)
69	Breeding for horizontal resistance to wheat diseases, 1986 (E)
70	Breeding for durable resistance in perennial crops, 1986 (E)
71	Technical guideline on seed potato micropropagation and multiplication, 1986 (E)
72/1	Pesticide residues in food 1985 – Evaluations – Part I: Residues, 1986 (E)
72/2	Pesticide residues in food 1985 – Evaluations – Part II: Toxicology, 1986 (E)
73	Early agrometeorological crop yield assessment, 1986 (E F S)
74	Ecology and control of perennial weeds in Latin America, 1986 (E S)
75	Technical guidelines for field variety trials, 1993 (E S)

76 Guidelines for seed exchange and plant introduction in tropical crops, 1986 (E)
77 Pesticide residues in food 1986 – Report, 1986 (E F S)
78 Pesticide residues in food 1986 – Evaluations – Part I: Residues, 1986 (E)
78/2 Pesticide residues in food 1986 – Evaluations – Part II: Toxicology, 1987 (E)
79 Tissue culture of selected tropical fruit plants, 1987 (E)
80 Improved weed management in the Near East, 1987 (E)
81 Weed science and weed control in Southeast Asia, 1987 (E)
82 Hybrid seed production of selected cereal, oil and vegetable crops, 1987 (E)
83 Litchi cultivation, 1989 (E S)
84 Pesticide residues in food 1987 – Report, 1987 (E F S)
85 Manual on the development and use of FAO specifications for plant protection products, 1987 (E F S)
86/1 Pesticide residues in food 1987 – Evaluations – Part I: Residues, 1988 (E)
86/2 Pesticide residues in food 1987 – Evaluations – Part II: Toxicology, 1988 (E)
87 Root and tuber crops, plantains and bananas in developing countries – challenges and opportunities, 1988 (E)
88 *Jessenia* and *Oenocarpus*: neotropical oil palms worthy of domestication, 1988 (E S)
89 Vegetable production under arid and semi-arid conditions in tropical Africa, 1988 (E F)
90 Protected cultivation in the Mediterranean climate, 1990 (E F)
91 Pastures and cattle under coconuts, 1988 (E)
92 Pesticide residues in food 1988 – Report, 1988 (E F S)
93/1 Pesticide residues in food 1988 – Evaluations – Part I: Residues, 1988 (E)
93/2 Pesticide residues in food 1988 – Evaluations – Part II: Toxicology, 1989 (E)
94 Utilization of genetic resources: suitable approaches, agronomical evaluation and use, 1989 (E)
95 Rodent pests and their control in the Near East, 1989 (E)
96 *Striga* – Improved management in Africa, 1989 (E)
97/1 Fodders for the Near East: alfalfa, 1989 (Ar E)
97/2 Fodders for the Near East: annual medic pastures, 1989 (Ar E F)
98 An annotated bibliography on rodent research in Latin America 1960-1985, 1989 (E)
99 Pesticide residues in food 1989 – Report, 1989 (E F S)
100 Pesticide residues in food 1989 – Evaluations – Part I: Residues, 1990 (E)
100/2 Pesticide residues in food 1989 – Evaluations – Part II: Toxicology, 1990 (E)
101 Soilless culture for horticultural crop production, 1990 (E)
102 Pesticide residues in food 1990 – Report , 1990 (E F S)
103/1 Pesticide residues in food 1990 – Evaluations – Part I: Residues, 1990 (E)
104 Major weeds of the Near East, 1991 (E)
105 Fundamentos teórico-prácticos del cultivo de tejidos vegetales, 1990 (S)
106 Technical guidelines for mushroom growing in the tropics, 1990 (E)
107 *Gynandropsis gynandra* (L.) Briq. – a tropical leafy vegetable – its cultivation and utilization, 1991 (E)
108 La carambola y su cultivo, 1991 (S)
109 Soil solarization, 1991 (E)
110 Potato production and consumption in developing countries, 1991 (E)
111 Pesticide residues in food 1991 – Report, 1991 (E)

112	Cocoa pest and disease management in Southeast Asia and Australasia, 1992 (E)
113/1	Pesticide residues in food 1991 – Evaluations – Part I: Residues, 1991 (E)
114	Integrated pest management for protected vegetable cultivation in the Near East, 1992 (E)
115	Olive pests and their control in the Near East, 1992 (E)
116	Pesticide residues in food 1992 – Report, 1993 (E F S)
117	Quality declared seed, 1993 (E)
118	Pesticide residues in food 1992 – Evaluations – Part I: Residues, 1993 (E)
119	Quarantine for seed, 1993 (E)

Availability: November 1993

Ar	–	Arabic	Multil	– Multilingual
C	–	Chinese	*	Out of print
E	–	English	**	In preparation
F	–	French		
P	–	Portuguese		
S	–	Spanish		

The FAO Technical Papers are available through the authorized FAO Sales Agents or directly from Distribution and Sales Section, FAO, Viale delle Terme di Caracalla, 00100 Rome, Italy.